MECHANICAL VIBRATIONS WITH APPLICATIONS

ELLIS HORWOOD SERIES IN MECHANICAL ENGINEERING

STRENGTH OF MATERIALS
Vol. 1: Fundamentals – Vol. 2: Applications
J. M. ALEXANDER, University College, Swansea
TECHNOLOGY OF ENGINEERING MANUFACTURE
J. M. ALEXANDER, University College, Swansea, G. W. ROWE, Birmingham University and R. C. BREWER
VIBRATION ANALYSIS AND CONTROL SYSTEM DYNAMICS
C. BEARDS, Imperial College of Science and Technology
STRUCTURAL VIBRATION ANALYSIS
C. BEARDS, Imperial College of Science and Technology
COMPUTER AIDED DESIGN AND MANUFACTURE 2nd Edition
C. B. BESANT, Imperial College of Science and Technology
BASIC LUBRICATION THEORY 3rd Edition
A. CAMERON, Imperial College of Science and Technology
SOUND AND SOURCES OF SOUND
A. P. DOWLING and J. E. FFOWCS-WILLIAMS, University of Cambridge
MECHANICAL FOUNDATIONS OF ENGINEERING SCIENCE
H. G. EDMUNDS, University of Exeter
ADVANCED MECHANICS OF MATERIALS 2nd Edition
Sir HUGH FORD, F.R.S., Imperial College of Science and Technology, and J. M. ALEXANDER, University College of Swansea.
MECHANICAL FOUNDATIONS OF ENGINEERING SCIENCE
H. G. EDMUNDS, Professor of Engineering Science, University of Exeter
ELASTICITY AND PLASTICITY IN ENGINEERING
Sir HUGH FORD, F.R.S. and R. T. FENNER, Imperial College of Science and Technology
DIESEL ENGINEERING PRINCIPLES AND PERFORMANCE
S. D. HADDAD, Associate Professor, Western Michigan University, USA, and Director of HTCS Co., UK, and N. WATSON, Reader in Mechanical Engineering, Imperial College of Science and Technology, University of London
DIESEL ENGINEERING DESIGN AND APPLICATIONS
S. D. HADDAD, Associate Professor, Western Michigan University, USA, and Director of HTCS Co., UK. and N. WATSON, Reader in Mechanical Engineering, Imperial College of Science and Technology, University of London
TECHNIQUES OF FINITE ELEMENTS
BRUCE M. IRONS, University of Calgary, and S. AHMAD, Bangladesh University, Dacca
FINITE ELEMENT PRIMER
BRUCE IRONS and N. SHRIVE, University of Calgary
CONTROL OF FLUID POWER: ANALYSIS AND DESIGN 2nd (Revised) Edition
D. McCLOY, Ulster Polytechnic, N. IRELAND and H. R. MARTIN, University of Waterloo, Ontario, Canada
UNSTEADY FLUID FLOW
R. PARKER, University College, Swansea
DYNAMICS OF MECHANICAL SYSTEMS 2nd Edition
J. M. PRENTIS, University of Cambridge
ENERGY METHODS IN VIBRATION ANALYSIS
T. H. RICHARDS, University of Aston Birmingham
ENERGY METHODS IN STRESS ANALYSIS: With Intro. to Finite Element Techniques
T. H. RICHARDS, University of Aston in Birmingham
COMPUTATIONAL METHODS IN STRUCTURAL AND CONTINUUM MECHANICS
C. T. F. ROSS, Portsmouth Polytechnic
FINITE ELEMENT PROGRAMS FOR AXISYMMETRIC PROBLEMS IN ENGINEERING
C. T. F. ROSS, Portsmouth Polytechnic
ENGINEERING DESIGN FOR PERFORMANCE
K. SHERWIN, Liverpool University
ROBOTS AND TELECHIRS
M. W. THRING, Queen Mary College, University of London
MECHANICAL VIBRATIONS WITH APPLICATIONS
A. C. WALSHAW, Professor Emeritus, The University of Aston in Birmingham

MECHANICAL VIBRATIONS WITH APPLICATIONS

A. C. WALSHAW, B.Sc., M.Sc., D.I.C., Ph.D., D.Sc.
Professor Emeritus, The University of Aston in Birmingham
Formerly Professor and Head of Department of Mechanical Engineering
and Dean of the Faculty of Engineering

ELLIS HORWOOD LIMITED
Publishers · Chichester
Halsted Press: a division of
JOHN WILEY & SONS
New York · Chichester · Brisbane · Toronto

First published in 1984 by

ELLIS HORWOOD LIMITED

Market Cross House, Cooper Street, Chichester, West Sussex, PO19 1EB, England

The publisher's colophon is reproduced from James Gillison's drawing of the ancient Market Cross, Chichester.

Distributors:

Australia, New Zealand, South-east Asia:
Jacaranda-Wiley Ltd., Jacaranda Press,
JOHN WILEY & SONS INC.,
G.P.O. Box 859, Brisbane, Queensland 40001, Australia

Canada:
JOHN WILEY & SONS CANADA LIMITED
22 Worcester Road, Rexdale, Ontario, Canada.

Europe, Africa:
JOHN WILEY & SONS LIMITED
Baffins Lane, Chichester, West Sussex, England.

North and South America and the rest of the world:
Halsted Press: a division of
JOHN WILEY & SONS
605 Third Avenue, New York, N.Y. 10016, U.S.A.

British Library Cataloguing in Publication Data
Walshaw, A. C.
Mechanical vibrations with applications. –
(Ellis Horwood series in mechanical engineering)
1. Vibration
I. Title
620.3 TA355

Library of Congress Card No. 84-12962

ISBN 0-85312-593-7 (Ellis Horwood Limited – Library Edn.)
ISBN 0-85312-628-3 (Ellis Horwood Limited – Student Edn.)
ISBN 0-470-20115-0 (Halsted Press)

Printed in Great Britain by R. J. Acford, Chichester.

Contents

Preface

The primary purpose of this book is to provide students in the early stages of degree and diploma courses with a gradual and straightforward approach to the subject in a compact volume of self-teaching material. Practising engineers, designers and draughtsmen will also find it useful as a refresher course leading to more specialized literature on vibration and noise reduction, control-system dynamics and computer-aided analysis.

Theoretical and practical aspects are inter-linked—the text being interspersed with worked examples which give practical significance to theory and analysis. Also, since the process of learning is considerably aided by problem-solving, exercises with answers are included at the ends of chapters.

Although a certain dexterity in mathematics is required for the elucidation of problems involving vibrations, only a knowledge of mechanics and calculus is required in accordance with their progressive acquisition in degree and diploma courses in engineering science. Thus, on a basis of 'A' level mathematics, the text begins with physical systems of simple oscillatory or harmonic motion, and gradually builds up via natural vibrations to damped forced vibratory systems and continuous-mass systems. In the earlier chapters emphasis is placed on single-degree-of-freedom systems—these being the most useful with which to introduce the subject. A parallel effort has been made throughout to reveal physical meanings and practical applications; and many cross-references inter-connect the Sections of the text.

The analysis and predictions of responses and stress in dynamic systems are often inherently more difficult than that of stressing in static systems—one difficulty being interpreting the physical significance of mathematical equations and solutions. Hence, visual aids in the form of diagrammatic sketches, vector and phase diagrams and graphs complementary to their mathematical counterparts, are included to facilitate understanding the physical meanings of equations and relationships between elastic, inertia, friction and applied forces in vibrating systems. Thus, regard has been given to students meeting the subject for the first time, and to those who may not be attracted to a wholly mathematical treatment. Proofs of formulae fundamental to mechanical vibration and dynamics of machines, and solutions to equations used in the text are summarized in Appendices for easy reference and confirmation of formulae often memorised wholly or in part.

Overall, the aim has been to present the subject-matter in as easy a manner as possible, consistent with accuracy, so that the study of mechanical vibrations can be pursued throughout degree and diploma courses in engineering science, and a base established from which complex dynamic systems can be investigated using computer-aided techniques in post-graduate studies and research.

My thanks are due to Dr J. E. Penny, Senior Lecturer in Mechanical Engineering with responsibility for vibrations and computer applications, for valuable comments and suggestions; and for reading the manuscript and proofs of the book.

The University of Aston in Birmingham, 1984 A.C.W.

Preliminaries

0.1 KEY TO THE MAIN USE OF SYMBOLS—FOR EASY REFERENCE

Letter-symbol	Signification	Unit-symbol (SI)
a	linear acceleration	m/s^2
$A, B, C \ldots$	constants of integration	
b	length	m
c	damping coefficient (force/velocity)	Ns/m
D, d	operator indicating differentiation	
d	static displacement or deflection	m
E	Young's modulus of elasticity	N/m^2
E, e	energy (a scalar product)	Nm = J
F, f	force (a vector quantity)	$N = kg\,m/s^2$
f	cyclic frequency	cycle/s
G	modulus of rigidity	N/m^2
g	acceleration of free fall	m/s^2
h	hysteretic damping constant	$N/m = kg/s^2$
I	moment of inertia (second moment of mass)	$kg\,m^2 = Nms^2$
J	second polar moment of area	m^4
j	square root of minus one; impulse	$\sqrt{(-1)}$; Ns
k	stiffness (force/displacement)	$N/m = kg/s^2$
k_O, k_G	radius of gyration	m
L, l	length	m
M, m	mass	kg
n	rotational frequency ($n = \omega/2\pi$)	$rev/s = 2\pi/s$
P, p	pressure; power	N/m^2 = Pa; W = J/s
R, r	radius	m
T	torque or twisting moment (a vector product)	N × m
	kinetic energy (a scalar product)	N·m = J

Symbol	Quantity	Unit
t	time	s
V	linear velocity; potential energy	m/s; Nm = J
v	linear velocity	m/s
W	weight ($W = mg$)	$N = kg\ m/s^2$
w	intensity of loading per unit length of beam or spring	N/m
X, x; Y, y	linear displacement	m
$\propto$	angular acceleration	$rad/s^2 = 1/s^2$
$\alpha, \beta, \gamma, \theta, \phi, \psi$	plane angles (non-dimensional)	rad
δ	logarithmic decrement ($\log_\varepsilon (x_1/x_2)$)	non-dimensional
ε	base of Naperian logarithms (2.71828....)	non-dimensional
ζ	damping factor or ratio ($c/c_c = c/(2\sqrt{km} = c\omega_n/2k$)	non-dimensional
η	coefficient of dynamic viscosity	kg/ms
Θ	amplitude of vibration with respect to θ co-ordinate	rad
λ	group of variables $\left(\lambda^4 = \dfrac{\rho a \omega^2}{EI}\right)$ in vibrating beams	1/m
μ	coefficient of friction	non-dimensional
π	ratio: circumference/diameter of a circle (3.1416...)	non-dimensional
ρ	density	kg/m^3
τ	periodic time	s
ω	angular velocity (rad/time), angular frequency	rad/s = 1/s
ω_n	angular frequency of undamped natural oscillations	rad/s = 1/s
ω_d	angular frequency of damped natural oscillations	rad/s = 1/s

0.2 SIGN CONVENTIONS AND FREE-BODY DIAGRAMS

The choice of co-ordinates (Cartesian or polar) to specify position is one of convenience with a view to easing the analysis involved, e.g. such that the weight of (gravitational force $W = mg$) a body gives a positive displacement (x) vertically downwards of, say, a helical spring which may be supporting a body of mass m (see Fig. 0.5(b)). Mass, being scalar quantity, has magnitude but is independent of direction; but, force, displacement, velocity and acceleration, being vector quantities, have direction as well as magnitude. So-called 'absolute' values are, of course, relative to the earth, and usually thus: ⊥ x → x showing the positive direction of displacement x, and as a reminder that, e.g. it is the 'absolute' acceleration ($\ddot{x} = a$) that must always be used in Newton's second law of motion ($F = ma$). In general, when formulating equations of motion, difficulties with signs (+ and −) can be avoided if displacements (x) velocities ($\dot{x}$) and accelerations ($\ddot{x}$) are always assumed, in the first instance, to be in the positive direction as arrowed (say, $\overset{x}{\rightarrow}$) which, for example, when a helical spring with a fixed end is involved, is usually (and best) taken to be in the direction of extension (rather than of compression) of the non-fixed end. In the case of torques (T) and angular displacements (θ), positive directions are usually taken to be in the anti-clockwise direction, in accordance with the customary convention in mathematics. Axes should *always* be drawn to indicate positive directions of vector quantities (e.g. x, $\dot{x}$, $\ddot{x}$; θ, $\dot{\theta}$, $\ddot{\theta}$; $F = m\ddot{x}$, $T = I\ddot{\theta}$) and great care taken in adhering to a consistent sign convention pre-supposed in a particular problem (see Figs. 0.5(b), 1.3(c), 1.4(b), 2.2(b), 2.3(e), 2.6(g), 3.2(a), 3.8(a), 4.4(a), 5.1(a), 6.1(a).)

Free-body diagrams enable any difficulties in determining the relationships between forces or torques and motion (displacements x, θ; velocities $\dot{x}$, $\dot{\theta}$; accelerations $\ddot{x}$, $\ddot{\theta}$) to be clarified and overcome by the simple procedure of:

(i) isolating the body (or system) within a boundary (real or imaginary) by drawing its 'free-body diagram'—thus clearly defining what is under consideration;
(ii) showing by arrowed vectors all forces (or torques) on that body or system;
(iii) clearly indicating the positive directions (x θ) of displacements.
(iv) equating the resultant force in the positive direction to mass x linear acceleration ($F = ma = m\ddot{x}$) and similarly, for torque ($T = I\alpha = I\ddot{\theta}$).

0.3 MANIPULATION AND REDUCTION OF UNITS

Conversion and reduction of units can be achieved with certainty if 'unity multipliers' are used in accordance with definitions, e.g. the SI unit of force, namely, $\mathrm{N} = \mathrm{kg\,m/s^2}$ yields $\left[\dfrac{\mathrm{Ns^2}}{\mathrm{kg\,m}}\right] = 1$ (unity).

Similarly,

$$\left[\frac{\pi\ \mathrm{rad}}{180\ \mathrm{deg}}\right], \left[\frac{60\ \mathrm{s}}{\mathrm{min}}\right], \left[\frac{2\pi\ \mathrm{rad}}{\mathrm{rev}}\right], \left[\frac{\mathrm{Nm}}{\mathrm{Ws}}\right],$$

etc., are each 'unity brackets' which, *since multiplying or dividing any quantity by unit* [1] *does not alter its amount*, can be inserted any time a conversion or reduction of units is required. For example, the pull of the earth on a body of mass m at a place where the acceleration of free fall is g is given by $W = mg$

which, of course, changes with the value of g. At Sèvres, so-called 'standard gravity' is $g_n = 9.80665\ \mathrm{m/s^2}$ hence, on the earth, the gravitational force or weight of one kilogram may be taken as

$$W = 1\,\mathrm{kg} \times 9.807\frac{\mathrm{m}}{\mathrm{s}^2}\left[\frac{\mathrm{Ns}^2}{\mathrm{kg\,m}}\right] = 9.807\,\mathrm{N},$$

using the 'Newtonian unity bracket' for conversion or reduction of the units. Although, in many cases, unit-symbols can safely be crossed out when replaced by their equivalents, errors are certainly avoided if appropriate 'unity brackets' are *written-in* before like-units in numerator and denominator are cancelled with each other. Also, *to have written them down enables subsequent checks to be made more easily* on quantitative and arithmetical work.

We may note here that 'radian', being the ratio of equal lengths, is the name of a plane angle whose 'circular measure' is unity. Hence, 'rad' can be replaced by unity (i.e. in effect omitted) in a calculation at any convenient point—it having been a help to write, say, $1\ \mathrm{rev} = 2\pi\ \mathrm{rad}$, and angular acceleration α as a number of $\mathrm{rad/s^2}$, etc, until the arithmetical stage of a problem has been reached, when rad can be replaced by unity, e.g. the energy transmitted by a rotating shaft is

$$W = T\theta = 6\,\mathrm{Nm} \times 30\,\mathrm{deg}\left[\frac{\pi\,\mathrm{rad}}{180\,\mathrm{deg}}\right] = \pi\,\mathrm{Nm} = \pi\mathrm{J}.$$

Because of the advantages (see Appendix B8) of presupposing circular (arc/radius) or radian measure of plane angle θ (and its derivatives $\dot{\theta} = \omega$ and $\ddot{\theta} = \alpha$) in algebraic analysis and resulting formulae, it should be understood, at the outset, that mathematicians and engineers tacitly assume that the values of plane angles will be expressed in radians. Thus, in all the usual formulae involving plane angle (θ) and its derivatives (ω and α) e.g. $S = r\theta$; $W = T\theta$; $v = r\omega$; $P = T\omega$; $E = \frac{1}{2}I\omega^2$; $T = I\alpha$; etc, plane angles are presupposed measured in radians, otherwise such formulae are not correct.

Although the circular measure of plane angle is dimensionless (i.e. may be represented by a number) it is necessary in quantitative work to insert rad as an indicator (or 'identity unit' answering the question 'number of what'?) to ensure that plane angles have been quantified correctly in accordance with the requirements of certain formulae. Thus if, because of practical expediency, angles and angular motion are first measured by means of degrees, complete turns or revolutions, conversion to radian measure can be achieved by using the 'unity multipliers' $\left[\frac{\pi\,\mathrm{rad}}{180\,\mathrm{deg}}\right] = 1 = \left[\frac{2\pi\,\mathrm{rad}}{\mathrm{rev}}\right]$ to be followed by replacing rad by unity in the final stage of a calculation.

EXAMPLES 0.3—illustrative of the use of 'unity multipliers' or 'unity brackets' in conversion and reduction of units

(i) In the simple case of a body of mass m rotating about a fixed point at a radius r with angular velocity ω, its instantaneous linear (or tangential) velocity is given by $\dot{x} = v = r\omega$ but *only if ω represents the number of radians per unit of time*. This restriction can, of course, be emphasized if need be (also see Appendix B1) by writing $|\omega| = \left(\dfrac{\omega}{\text{rad}}\right)$, where $|\omega|$ denotes a number or numeric per unit of time.

(ii) $W = T\theta$ may be written $W = T\left(\dfrac{\theta}{\text{rad}}\right) = 6\,\text{Nm} \times \left(\dfrac{30\,\text{deg}}{\text{rad}}\right)\left[\dfrac{\pi\text{rad}}{180\,\text{deg}}\right] = \pi\,\text{Nm}$

Alternatively, $W = T\theta = 6\,\text{Nm} \times \dfrac{\pi}{6}\,\text{rad} = \pi\,\text{Nm} = \pi\text{J}$, i.e. replacing rad by 1.

(iii) Similarly, $T = I\left(\dfrac{\alpha}{\text{rad}}\right) = 10\,\text{kg}\,\text{m}^2\left(\dfrac{2\ \text{rad}}{\text{rad s}^2}\right)\left[\dfrac{\text{N}\,\text{s}^2}{\text{kg}\,\text{m}}\right] = 20\,\text{Nm}$ or, more usually,

$$T = I\alpha = 10\,\text{kg}\,\text{m}^2\left(2\frac{\text{rad}}{\text{s}^2}\right)\left[\frac{\text{N}\,\text{s}^2}{\text{kg}\,\text{m}}\right] = 20\,\text{Nm}$$

(iv) The kinetic energy of the wheel of a gyroscope of moment of inertia $I = 1200$ kg mm^2 (or 1.2 mN ms^2) spinning at the rate

$$\omega = 20\,000\ \text{rev/min is}\ \tfrac{1}{2}I\omega^2 = \frac{1200}{2}\,\text{kg mm}^2\left(2 \times 10^4\,\frac{\text{rev}}{\text{min}}\right)^2$$

$$= \frac{24 \times 10^{10}}{36 \times 10^2}\,\text{kg mm}^2\,\frac{\text{rev}^2}{\text{s}^2}\left[\frac{\text{m}^2}{10^6\,\text{mm}^2}\right]\left[\frac{4\pi^2}{\text{rev}^2}\right]\left[\frac{\text{Ns}^2}{\text{kgm}}\right]$$

$$= 2.63 \times 10^3\ \text{Nm or } 2.63\ \text{kJ}.$$

These easy examples show that the writing in and cancellation of like-units, together with replacing radian (rad) by unity, provides an almost 'foolproof' or self-checking process in quantitative work so far as units are concerned.

0.4 FREQUENCY AND ITS MEASUREMENT

In engineering, customary ways of specifying frequencies of wave forms, rotating shafts and rotating vectors, harmonic and oscillatory motions are by measuring:

(i) rotational frequency (n) or rotational rate (rev/s),
(ii) angular (or circular) frequency (ω) or angle rate (rad/s), or
(iii) cyclic frequency (f) or cyclic rate (cycle/s).

For harmonics these are inter-related because 1 rev = 1 cycle = 2π rad—thinking of a uniformly rotating vector (see Fig. 0.5(a)), crank or arm. Hence, using these modes of measurement of the physical quantity (expressed by a number and its unit) designated by the generic word 'frequency' and represented by, say, the letter-symbol q whose periodic time is $\tau = 1$ cycle/q, we may write:

$$\frac{q}{2\pi\ \text{rad/s}} = \frac{q}{\text{rev/s}} = \frac{q}{\text{cycle/s}}\ \text{and}\ \frac{\tau}{s} = \frac{\text{cycle/s}}{q}.$$

This may, alternatively, be written:

$$\frac{q}{\text{rad/s}} = 2\pi\left(\frac{q}{\text{rev/s}}\right) = 2\pi\left(\frac{q}{\text{cycle/s}}\right) \text{ and } \frac{\tau}{\text{s}} = 2\pi\left(\frac{\text{rad/s}}{q}\right)$$

or

$$|\omega| = 2\pi|n| = 2\pi|f| \text{ and } |\tau| = \frac{2\pi}{|\omega|} = \frac{1}{|f|},$$

where $q = |\omega|$ rad/s $= |n|$ rev/s $= |f|$ cycle/s and $\tau = |\tau|$s, i.e. $|\omega|$, $|n|$ and $|f|$ are the numerical values of the same physical quantity, q, measured in different ways—the ratio of the numbers or numerics being $\dfrac{|\omega|}{|f|} = 2\pi = \dfrac{|\omega|}{|n|}$. Thus, we see that the familiar equations (usually written without the indicator $|\ |$ signifying numerics) $\omega = 2\pi n = 2\pi f$ and $\tau = \dfrac{1}{f} = \dfrac{2\pi}{\omega}$ are numerical equations necessarily implying (i.e. restricted to) the customary ways of measurement of which only s (the base unit of time) and rad (the SI 'supplementary unit' for plane angle) are units of the SI.

It is, however, always necessary (as, e.g. in Example 0.5(b)) to write rev, cycle, rad alongside their respective numerical values in calculations for eventual reduction to rad, which is dimensionless and can be replaced by unity (i.e. in effect omitted) at any convenient point in the final stages of a calculation as in Example 0.3 (also see Appendix B8).

Special care is needed in the use of the unit named 'hertz' (Hz) which is defined in general terms in the SI as 'the frequency of a periodic phenomenon of which the period is one second' and listed as Hz = 1/s. Hence, *it is necessary when* Hz *is used to state precisely what it is that recurs each second*, e.g. a heart-pulse of periodic time 0.75s may be said to have a frequency of 1.33 Hz; and in specifying a frequency as, say, 60 Hz it is necessary that there be no doubt as to what it is that recurs 60 times a second.

In this book 1Hz means 1 cycle/s—i.e. the unit-symbol, Hz, represents the cyclic frequency of harmonics inter-related by 1 rev = 1 cycle = 2π rad = 2π = Hz s. Hence, for harmonics (visualized, say, as tracing a sinusoidal wave (as in Fig. 0.5(a)) or one cycle each revolution of a rotating vector with constant angular velocity), the unity multiplier for conversion of units and customary modes of measurement are $\left[\dfrac{2\pi\text{ rad}}{\text{Hz s}}\right], \left[\dfrac{\text{rev}}{2\pi\text{ rad}}\right], \left[\dfrac{\text{Hz s}}{\text{cycle}}\right]$. For example, a crank or vector rotating with a constant angular velocity of magnitude 6π rad/s may alternatively be expressed as having a frequency $\omega = 6\pi\dfrac{\text{rad}}{\text{s}}\left[\dfrac{\text{Hz s}}{2\pi\text{ rad}}\right] = 3$ Hz or 3 rev/s or 3 cycle/s. This, of course, complies with the previous numerical formula $\omega = 2\pi n = 2\pi f$ and shows that algebraic analysis of problems involving frequencies can be carried out using one letter-symbol, ω, representing angular velocity (or angular frequency) which is easily convertible into either of the other ways (rotational or cyclic) of expressing frequency deemed appropriate in a particular case (see Example 0.5).

0.5 SIMPLE HARMONIC MOTION

S.h.m. warrants detailed analysis at the outset since it is of fundamental importance in engineering. Problems involving complicated vibrating systems are usually analysed, in the first instance, by replacing actual systems with simpler idealized or equivalent systems of block masses, assuming massless springs and damping devices—i.e. by simpler models in order to enable solutions to be found by mathematical analysis. Although the resulting solutions only approximate to reality, measurements on experimental analogous systems—electrical (see Appendix C9) as well as mechanical, may suggest modifications which will enable the performance of an actual system to be predicted, or vibrations to be eliminated if desired. Actual vibrations can, of course, only be detected and measured with the aid of recording instruments (see Appendix C) placed in appropriate places on real systems. *Engineers, and scientists generally, are never absolutely sure that any prediction of theory is correct until it has been tested and verified in practice.* Hence the importance of experimental and practical work alongside theory and analysis.

By definition of s.h.m., acceleration ($\ddot{x}$) is negatively proportional to displacement (x) in rectilinear oscillatory motion, i.e. $\ddot{x} \propto (-x)$ which can, alternatively, be expressed as

$$\ddot{x} = -\omega^2 x \text{ or } \frac{dx^2}{dt^2} + \omega^2 x = 0,$$

where ω is a constant as yet unidentified. This equation is satisfied by $x = A \sin(\omega t + \beta)$ as can easily be verified, where A and β are arbitrary constants in the solution of the second order (also see Appendix B 11 and 12) differential equation.

It follows that s.h.m., as a sine (or cosine) function of time, can also be represented by the varying length of the diametral projection (x) of the tip of a vector of length A rotating on a circular path with constant angular velocity (ω), necessarily expressed in radians per unit of time because this pre-supposition is implicit in the trigonometrical expression $x = A \sin(\omega t + \beta)$ in which β is a plane angle in radian measure, and A is a constant length called the amplitude (see Fig. 0.5(a) and also Section 1.6).

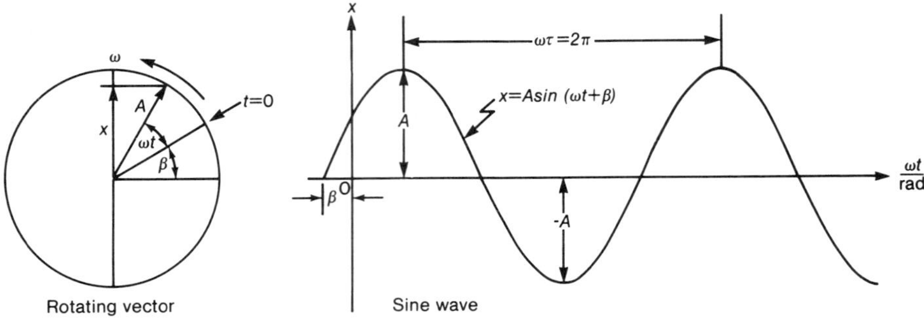

Fig. 0.5(a) Diagrammatic of s.h.m.

EXAMPLE 0.5—illustrative of the advantage of using angular velocity or angular frequency (ω) in mechanical vibration analysis

(a) Analyse the relatively simple case of a mass m oscillating freely (i.e. in undamped natural oscillation) in rectilinear (vertical) motion on the end of a helical spring of elastic stiffness k (i.e. representing force per unit extension). Show that the oscillation is s.h.m., and deduce expressions for angular frequency (ω) and periodic time (τ).

(b) An instrument of mass $m = 10$ kg is set on three rubber mountings each of which has a deflection rating of 0.75 mm/N. Calculate the cyclic frequency (f), the periodic time (τ) of the oscillations and the angular or circular frequency (ω) of a corresponding rotating vector. Neglect the effect of the mass of the spring and any damping.

(a) Referring to Fig. 0.5(b), if the static extension of the loaded spring is d, then the equation of motion, in accordance with Newton's second law $F = ma$ when the mass m is displace a distance x from its static-equilibrium position is $mg - k(d + x) = m\ddot{x}$ as deduced from the free-body diagram. Alternatively, since $mg = kd$, then $\dfrac{\mathrm{d}^2x}{\mathrm{d}t^2} = -\left(\dfrac{k}{m}\right)x$ which is an equation for s.h.m.

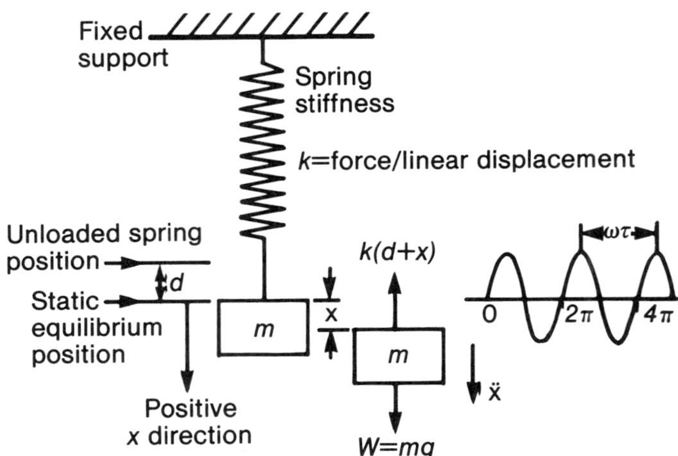

Fig. 0.5(b) A mass in s.h.m.

By trial, $x = A\sin\omega t + B\cos\omega t = X\sin(\omega t + \phi)$ proves to be a solution because $\dfrac{d^2x}{dt^2} = -\omega^2 x$. Also, $X = \sqrt{(A^2 + B^2)}$ and $\phi = \tan^{-1}(B/A)$ and $\omega^2 = k/m$.

If the starting or initial conditions are $t = 0$, $x = X$, $\dot{x} = 0$, then $X = X\sin\phi$ and $\dot{x} = 0 = \omega X\cos\phi$; hence $\phi = \pi/2$ rad and $x = X\sin(\omega t = \pi/2) = X\cos\omega t$.

It follows that if $x = X\sin(\omega t + \phi)$ is to represent the s.h.m., then $\omega = \sqrt{(k/m)}$, which has dimensions $[\tau^{-1}]$. In emphasis of the fact that such mathematical solutions presuppose plane angle (θ) and its derivative ($\dot{\theta} = \omega$) to be in circular measure (i.e. ω is, necessarily, expressed in radius perunit of time) this result may be written more explicity as

$$\frac{\dot{\theta}}{\text{rad}} = \frac{\omega}{\text{rad}} = \sqrt{(k/m)} = \sqrt{(g/d)}.\ \text{Hence, } \omega = \{\sqrt{(k/m)}\}\,\text{rad}.$$

The cycle of motion is completed at intervals whose periodic time (τ) is given by the physical equations $f\tau = 1$ cycle or $\omega\tau = 2\pi$ rad, where f and ω are the cyclic and angular frequencies, respectively.

(b) The stiffness of the three rubber mountings taken together is equivalent to $k = \dfrac{3\text{N}}{0.75\,\text{mm}} = 4\text{kN/m}$. Hence, the angular frequency of the system is $\omega = \{\sqrt{(k/m)}\}\,\text{rad} = \left\{\sqrt{\left(4\times 10^3\dfrac{\text{N}}{\text{m}} \times \dfrac{1}{10\,\text{kg}}\left[\dfrac{\text{kg m}}{\text{N s}^2}\right]\right)}\right\}\text{rad} = 20\,\text{rad/s}$ which, since for s.h.m. $\left[\dfrac{\text{cycle}}{2\pi\,\text{rad}}\right] = 1$, converts to a cyclic frequency $f = 3.18$ cycle/s or 3.18 Hz; and the periodic time is given by the physical equation

$$\tau = \frac{2\pi}{\omega}\text{rad} = \frac{\text{cycle}}{f} = \frac{1\,\text{s}}{3.18} = 0.314\,\text{s}.$$

We may note that, since the magnitude of the same frequency relates angular and cyclic frequencies by the numerical factor 2π, it is unnecessary to use two symbols ω and f for frequency provided that the modes of measurement are written along side their corresponding numerical values. Thus, it will be appreciated that by using angular frequency (ω) rather than cyclic frequency (f, which is numerically equal to $\omega/2\pi$) we avoid a plethora of 2π's and $4\pi^2$'s in analysis or predictions of harmonic oscillations and vibrations.

1

Types of vibration; and relevant mechanics

1.1 TYPES OF VIBRATION

All matter—solid, liquid and gaseous—is capable of vibration, e.g. vibration of gases occurs in tail ducts of jet engines causing troublesome noise and sometimes fatigue cracks in the metal. Vibration in liquids is almost always longitudinal and can cause large forces because of the low compressibility of liquids, e.g. pipes conveying water can be subjected to high inertia forces (or 'water hammer') when a valve or tap is suddenly closed. Excitation forces caused, say, by changes in flow of fluids or out-of-balance rotating or reciprocating parts, can often be reduced by attention to design and manufacturing details. A typical machine has many moving parts, each of which is a potential source of vibration or shock-excitation. Designers face the problem of compromising between an acceptable amount of vibration and noise, and costs involved in reducing excitation.

In this book consideration is given to commonly-occurring sources of vibration; and methods of analysis are set forth to indicate ways of reducing the dynamic forces responsible for vibration and shock. Even then, relatively small excitation forces can cause large responses if near a resonant condition (see Section 1.2).

Practically every object or system vibrates freely on its own after some initial disturbance. Mechanical excitation is a necessity for the production of sound—'musical' or 'noise'—the audible range being about 18 Hz to 18 kHz. All musical instruments use vibrations caused by different excitations, e.g. drums and cymbals, are struck, and harp and guitar strings are plucked—i.e. impulse excitation. Violin and cello strings are bowed—i.e. 'stick-slip' excitation; wind instruments and organ pipes are blown—i.e. wind excitation. In contrast, whilst it is not easy to detect and measure the motion of, say, an engine part vibrating with an amplitude of a fraction of a millimetre, it is relatively easy in the case of musical instruments.

Engineers use vibrations deliberately in such operations as the drilling and cutting of square holes, etc, but in most cases the source and cause of vibration is sought in order to reduce or eliminate it if possible, otherwise high local stresses may be induced causing eventual failure of material by fatigue. Hence, the importance of periodic inspection of parts (e.g. crack-detection) where fatigue is a possibility.

Fortunately, the sine (or cosine) wave (see Fig. 1.3(a)) can be the basis of analysis of vibrations and oscillatory quantities because, however complicated or peculiar the shape of a periodic wave-form, it can always be built up as an assembly of suitable sine waves (see the Fourier series in Section 1.3).

The mechanical vibrations dealt with are either excited by steady harmonic forces (i.e. obeying sine and cosine laws in cases of *forced* vibrations) or, after an initial disturbance, by no external force apart from gravitational force called weight (i.e. in cases of *natural*, autonomous or free vibrations). Harmonic vibrations are said to be 'simple' (see Example 0.5) if there is only one frequency as represented diagrammatically by a sine or cosine wave of displacement against time, as in Figs. 0.5(a) and 1.5(a).

Vibration of a body or material is periodic change in position or displacement from a *static-equilibrium position* (see Fig. 0.5(b)). Associated with vibration are the inter-related physical quantities of acceleration, velocity and displacement (see Fig. 1.5(a))—e.g. an unbalanced force causes acceleration ($a = F/m$) in a system which, by resisting, induces vibration as a response. We shall see that vibratory or oscillatory motion may be classified broadly as (i) transient, (ii) continuing or steady-state, and (iii) random.

Transient vibrations die away and are usually associated with irregular disturbances, e.g. shock or impact forces, rolling loads over bridges, cars driven over pot holes, aircraft buffeted in air pockets—i.e. forces which do not repeat at regular intervals. Although transients are temporary components of vibrational motion, they can cause large amplitudes initially and consequent high stress but, in many cases, they are of short duration and can be ignored leaving only steady-state vibrations to be considered.

Steady-state vibrations are often associated with the continuous operation of machinery and, although periodic, are not necessarily harmonic or sinusoidal. Since vibrations require energy to produce them, they reduce the efficiency of machines and mechanisms because of dissipation of energy, e.g. by friction and consequent heat-transfer to surroundings, sound waves and noise, stress waves through frames and foundations, etc. Thus, steady-state vibrations always require a continuous energy-input to maintain them.

Random vibration is the term used for vibration which is not periodic, i.e. has no cyclic basis and is not regularly repetitive, e.g. vibration induced by wind gusts and vortices. Consequently, the usual methods of analysis using harmonics cannot be applied, but analysis may be attempted using statistical methods and computer programming beyond the range of this book.

As we shall see in succeeding Sections, vibrations can be reduced by removing or counterbalancing the periodic or exciting forces, e.g. by damping, 'detuning' or changing the natural frequency of the vibrating system, by other neutralizing devices (see Fig. 4.7(a)) or by a combination of such methods.

1.2 DEFINITIONS AND TERMINOLOGY

In this Section certain terms and definitions used in vibration analysis are made clear—several of which are probably known to science students already. For example, if a garden swing is pulled back and let go it will oscillate with its *natural frequency* in accordance with its design, and under the influence of gravity—the amplitude gradually reducing owing to air resistance. If, however, the swing is forced rythmically to and fro, the frequency will be that of the applied force; and if it is pushed every time it reaches the end of a natural cycle, the amplitude will build up because the frequency of the applied force is the same as the natural frequency of the swing. This is a case of *resonance* which, in general, is dangerous and, in machines, can cause high stresses and breakages.

Period, cycle, frequency and amplitude

A steady-state mechanical vibration is the motion of a system repeated after an interval of time known as the *period* (see Figs. 1.3(a) and 2.3(b)). The motion completed in any one period of time is called *a cycle*. The number of cycles per unit of time is called the *frequency* (see Section 0.4). The maximum displacement of any part of the system from its static-equilibrium position is the *amplitude* of the vibration of that part—the total travel being twice the amplitude. Thus, 'amplitude' is not synonymous with 'displacement' but is the maximum value of the displacement from the static-equilibrium position.

Natural and forced vibration

A natural vibration occurs without any external force except gravity (see Figs. 0.5(b) and 2.2(a)), and normally arises when an elastic system is displaced from a position of stable equilibrium and released, i.e. natural vibration occurs under the action of restoring forces inherent in an elastic system, and natural frequency is a *property* of the system.

A forced vibration takes place under the excitation of an external force (or externally-applied oscillatory disturbance) which is usually a function of time, e.g. in unbalanced rotating parts, imperfections in manufacture of gears and drives. The frequency of forced vibration is that of the exciting or impressed force, i.e. the forcing frequency is an arbitrary quantity independent of the natural frequency of the system.

Resonance

'Resonance describes the condition of maximum amplitude (see Fig. 4.1(g)). It occurs when the frequency of an impressed force coincides with, or is near to a natural frequency of the system (see Figs. 4.1(b) and 4.4(f)). In this critical condition, dangerously large amplitudes and stresses may occur in mechanical systems but, electrically, radio and television receivers are designed to respond to resonant frequencies. The calculation or estimation of natural frequencies is, therefore, of great importance in all types of vibrating and oscillating systems. When resonance occurs in rotating shafts and spindles, the speed of rotation is known as the *critical speed* (as analysed and predicted in Section 5.6). Hence, the prediction and correction or avoidance of a resonant condition in mechanisms is of vital importance since, in the absence of damping or other amplitude-limiting devices, resonance is the condition at which a system gives an infinite response (see Fig. 4.1(b)) to a finite excitation.

Damping

Damping is the dissipation of energy from a vibrating system, and thus prevents excessive response. It is observed that a natural vibration diminishes in amplitude with time and, hence, eventually ceases owing to some restraining or damping influence. Thus, if a vibration is to be sustained, the energy dissipated by damping must be replaced from an external source.

The dissipation is related in some way to the relative motion between the components or elements of the system, and is caused by frictional resistance of some sort, e.g. in structures, internal friction (or 'hysteretic' damping—see Section 3.8) in material, and external friction caused by air or fluid resistance called 'viscous' damping (see Section 3.1) if the drag force is assumed proportional to the relative velocity between moving parts. One device assumed to give viscous damping is the 'dashpot' which is a loosely-fitting piston (see Fig. 3.2(a)) in a cylinder so that fluid can flow from one side of the piston to the other through the annular clearance space. A dashpot cannot store energy but can only dissipate it.

Degress of freedom

The number of independent co-ordinates necessary to describe the motion of a system (e.g. see Section 2.2 and Fig. 5.1(a)) is equal to the number of degrees of freedom. Thus, if the motion of a body is constrained to vibrate or oscillate in only one manner or mode, it is said to have a single degree of freedom, e.g. a pendulum, or a mass oscillating vertically on the end of a helical spring (see Figs. 0.5(b) and 2.2(a)), whereas a body moving in space (e.g. a cricket ball or a satellite), or any unconstrained mass, has six degrees of freedom—3 translational and 3 rotational. Whilst most real systems have several degrees of freedom, simplifications can usually be made such that approximate systems with only one or two degrees of freedom can yield useful results and give valuable pointers as to what is happening in dynamically more complex systems. The general usefulness of such results is a strong incentive to use single-degree-of-freedom models wherever possible. As the number of degrees of freedom increases there is difficulty with correspondingly large numbers of variables, and it becomes impractical to present the results in general forms.

Principal modes of vibration

The word 'mode' relates to the shape or form of motion, e.g. linear or translational, angular or torsional, flexural, transverse or lateral. Thus, in the case of a string or wire stretched between two points, its simplest mode of vibration is when the middle of the string vibrates with maximum amplitude—called the fundamental mode. If, however, the wire is held at mid-span, the two halves will vibrate with twice the fundamental frequency (see Fig. 1.3(a)). Similarly, the string or wire can vibrate in three, four or moreparts at correspondingly higher frequencies. In such systems capable of natural vibration in a complex manner we shall see that there are simple orderly motions called principal modes of vibration which correspond to each natural frequency (see Section 5.8).

A characteristic of a principal mode is that all particles of a system pass through maximum and minimum (zero) velocities at the same time (see Fig. 5.4(b)). If the form of a principal mode is known, the corresponding expressions for (T) the kinetic and (V) the potential energies can be written down, and the natural frequency found relatively easily by equating T_{max} to V_{max} (see Examples in Section 2.4).

Nodes

The word 'node' applies to any point or line which is stationary at all times (e.g. the pivot line or axle of a see-saw) in a vibrating or oscillating system. Thus, if a point (or a cross-section in the case of torsion) does not move during vibration or oscillation of a system in one of the principal modes, i.e. the point or cross-section has zero amplitude, then a *node* is said to exist there (see Fig. 2.6(b)).

Control of vibrations

From a practical and design point of view, mechanical vibrations can be reduced or controlled by several techniques, as we shall see in subsequent sections, e.g.

(1) control of natural frequency to prevent resonance with an excitation frequency (as in Section 4.1).
(2) energy dissipation by damping devices (see Fig. 3.3(e) & Section 5.2),

(3) isolators to reduce transmission of excitation forces (see Section 4.6), and
(4) auxiliary-mass neutralizers (see Fig. 5.1(a)).

In some cases, however, the excitation frequency changes with operating requirements of machines and, hence, coincidence with natural frequency may be unavoidable at certain points in the working range. Similarly, if there is a broad band of random vibrations, resonance cannot be avoided but may be reduced by damping. In the case of shafts, the torque may not be steady but fluctuate because of successive pulses, say, by the intermittent firing in multi-cylinder engines and the rotation of cranks, gear-changing, etc. Hence, shaft design is difficult in such cases because a natural frequency may coincide with an excitation frequency at certain speeds of rotation.

1.3 PERIODIC MOTION

By definition, periodic motion repeats itself in regular intervals of time. *All harmonic motion is periodic, but not all periodic motion is harmonic* (see Example 1.3(ii). Waveforms are the graphs (or signatures) of cyclic variations against time as displayed, for example, on a cathode-ray oscilloscope. They can vary from a simple sine wave to complicated waveforms resulting from the combined effect of the parts of a system vibrating with different frequencies, e.g. the simple case of the sum of two harmonic terms $y = y_1 \sin \omega t + y_2 \sin 2\omega t$ gives the periodic $(y, \omega t)$ graph shown in Fig. 1.3(a). The first term has the same periodic time $(\tau = 2\pi/\omega)$ and, therefore, cyclic frequency $(f = \omega/2\pi = 1/\tau)$ as that of the function y, and is called the first harmonic or fundamental harmonic. The second term has twice the fundamental frequency, and is called the second harmonic. We thus see that *when harmonic motions of different frequencies are added together, the resulting periodic*

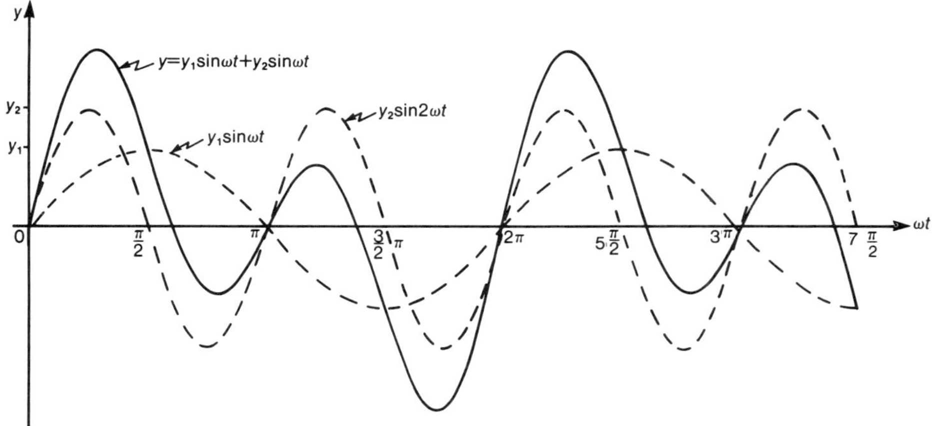

Fig. 1.3(a) Harmonic and periodic motion

motion is no longer harmonic. In other words, the vector sum of two vectors rotating with different frequencies does not remain of constant length as it rotates—i.e. the resulting Fig. 1.3(a) graph is not sinusoidal.

In mechanical engineering, periodic motion in the form of vibration (often caused by imbalance of rotating parts) is the cause of noise radiated from structural parts which sometimes act as resonating amplifiers. Many mechanisms have complex periodic motions which, fortunately, can be resolved into component s.h.m. waveforms; and there is often a dominant vibration of cyclic frequency (f) associated with an angular frequency (ω) or rotational frequency (n), where $\omega = 2\pi f = 2\pi n$ (see Section 0.4).

Fourier proved (in 1822) that all periodic waveforms (smooth or otherwise), e.g. those representing displacement or force as a function of time, can be resolved into a series of sine and cosine terms having frequencies which are multiples of a fundamental frequency ($\omega = \theta/t$). For example, the Fourier series for the rectangular wave form shown in Fig. 1.3(b) can be expressed as $y = \frac{4}{\pi}\left(\sin\theta + \frac{1}{3}\sin 3\theta + \frac{1}{5}\sin 5\theta + \cdots\right)$, and that for the periodic triangular form as $y = \frac{1}{2} + \frac{4}{\pi^2}\left(\cos\theta + \frac{1}{3^2}\cos 3\theta + \frac{1}{5^2}\cos 5\theta + \cdots\right)$.

Both the nth sine harmonic and the nth cosine harmonic, of course, go through n complete cycles in the fundamental period of the function y.

Fourier's theorem virtually states that any mechanical disturbance can be represented as a sum of harmonic terms or a combination of simple harmonic motions of different frequencies and amplitudes (also, see Section 5.2 and Appendix B6, B9 and D(9)). Such a series is very useful in analysis of periodic functions that are not harmonic, and which cannot be described or represented by a single sinusoidal function. Similarly, in cases where systems are excited by non-harmonic forces (e.g. to cause the motion of masses in the slider crank mechanism of Example 1.3(ii), and the torque from a reciprocating-piston engine), any force which varies with time in such a way that its value at any instant, t, is equal to its value at a preceding instant ($t - \tau$), where τ is fixed by, say, the

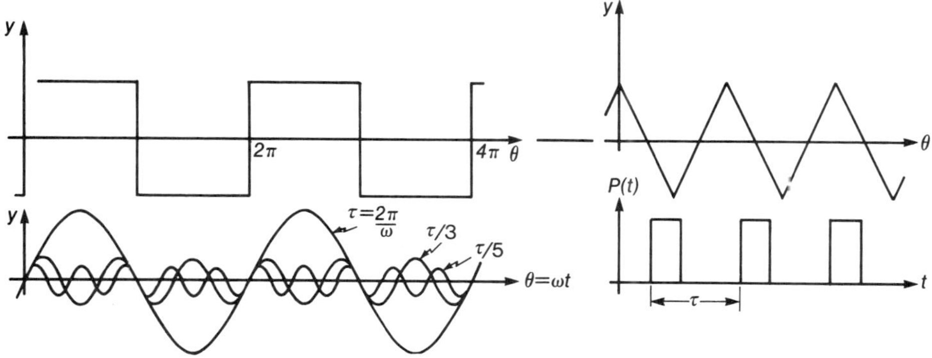

Fig. 1.3(b) Wave forms and pulses

running speed of a machine, may be expressed as a series of simusoidal components or harmonic terms with periods τ, $\tau/2$, $\tau/3$, etc. in conformity with Fourier's theorem.

Thus, the important result is that periodic forces of *any* waveform can be represented by a Fourier series as a superposition of harmonic components of various frequencies, in Fig. 1.3(b). The process of finding the sinusoidal components (harmonics) of a complex periodic function can be done relatively quickly by means of a spectrum analyser—a device which calculates the force coefficients digitally. Fortunately, in many problem, only equal frequency (i.e. the same ω) components are involved concerning elastic, inertia, damping and applied forces, which determine the shapes of phase diagrams (see Figs. 1.5(a), 2.3(a), 3.3(d), 3.3(f), 4.4(c)) as graphical representations of equations of instantaneous dynamic forces.

EXAMPLE 1.3(i)—illustrative of a linear-rotary system in s.h.m.
Deduce a formula for the natural frequency of small oscillations of the system shown diagrammatically in Fig. 1.3(c).

Using Newton's second law of motion ($F = m a$ and $T = I\alpha$ for translational and rotational motion, respectively) the equations of motion for the mass and cylinder, respectively, (as deduced from the free-body diagrams on the right of Fig. 1.3(c) are $W - F_w = m\alpha r$ and $F_w r - F_s b = J_0\alpha$ (also, see Appendix A1 & 2).
Hence, eliminating F_w, we get
$(W - m\alpha r)r - F_s b = J_0\alpha$ or, since $Wr = k\theta_0 b^2$ at the static-equilibrium position,

$$kb^2\theta_0 - m\alpha r^2 = k(\theta_0 + \theta)b^2 + J_0\alpha$$

which reduces to $(J_0 + mr^2)\alpha + kb^2\theta = 0$
or

$$\frac{d^2\theta}{dt^2} + \left(\frac{kb^2}{J_0 + mr^2}\right)\theta = 0 = \frac{d^2\theta}{dt^2} + \omega_n^2\theta,$$

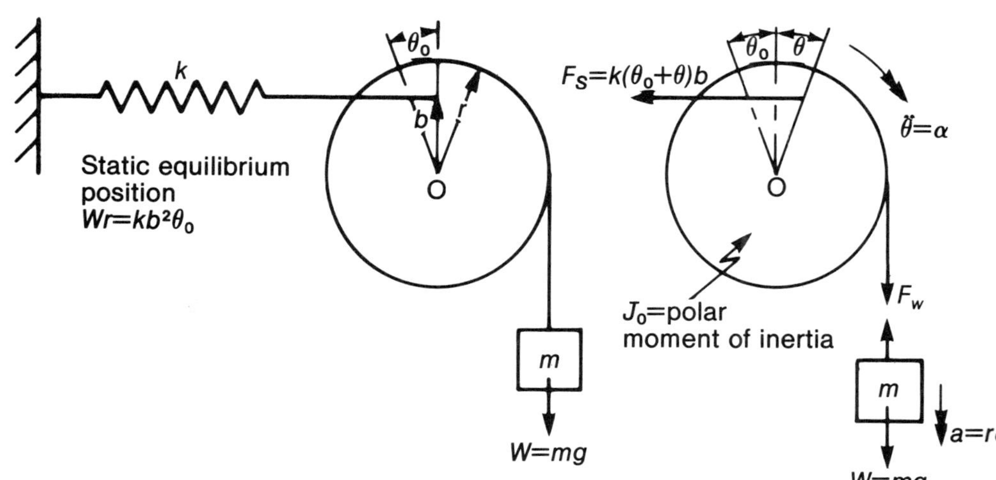

Fig. 1.3(c) A system in s.h.m.

where ω_n is the natural frequency (in radians per unit time) of the system in simple harmonic motion, namely, $\frac{\omega_n}{\text{rad}} = \sqrt{\left(\frac{kb^2}{J_0 + mr^2}\right)}$. The corresponding cyclic frequency f, is given by $\frac{f}{\text{cycle}} = \frac{\omega_n}{2\pi}$, i.e. the same frequency can be expressed in different modes of measurement, namely, radians/s and cycles/s whose ratio is $2\pi = 6.28$ (also, see Section 0.4).

EXAMPLE 1.3(ii)—illustrative of the fact that periodic motion is not, necessarily, harmonic.

(a) Show that the motion of the piston actuated by a crank and connecting rod is periodic but not simple harmonic. (b) Calculate the maximum acceleration and maximum inertia force on a piston of mass 1 kg linked to a connecting rod 100 mm long and crank of 40 mm which rotates at 3000 rev/min.

(a) Referring to Fig. 1.3(d) $y = r + l - l\cos\phi - r\cos\theta$.

Also, $l\sin\phi = r\sin\theta$ or $\sin\phi = R\sin\theta$, where $R = r/l$ and $\theta = \omega t$. Hence, since $l\cos\phi = l\sqrt{(1-\sin^2\phi)} = l(1 - R^2\sin^2\theta)^{1/2}$ which, by the binomial theorem (see Appendix B6) $l(1 - \frac{1}{2}R^2\sin^2\theta - \frac{1}{8}R^4\sin^4\theta - \frac{1}{16}R^6\sin^6\theta \ldots$, and, since $\sin^2\theta = \frac{1}{2}(1 - \cos 2\theta)$ and $\cos^2\theta = \frac{1}{2}(\cos 2\theta + 1)$—(see Appendix B5) so that $\sin^4\theta = \frac{1}{4}(1 - \cos 2\theta)^2 = \frac{1}{4}(\frac{3}{2} - 2\cos 2\theta + \frac{1}{2}\cos 4\theta)$ and $\sin^6\theta = \cdots$, etc., then $y = a_0 + a_1\cos\omega t + a_2\cos 2\omega t + a_3\cos 3\omega t + a_4\cos 4\omega t + \cdots$.
where

$$a_0 = r(1 + \tfrac{1}{2}R + \tfrac{3}{64}R^3 + \tfrac{5}{256}R^5 + \cdots);$$

$$a_1 = -r; a_2 = -r(\tfrac{1}{4}R + \tfrac{1}{16}R^3 + \tfrac{15}{512}R^3 + \cdots);$$

$$a_4 = r(\tfrac{1}{64}R^3 + \tfrac{1}{256}R^5 + \cdots);\text{ etc.}$$

Thus, the periodic displacement (y) of the piston (P) is not harmonic. If however $R = r/l$ is small enough to be neglected then $y = r - r\cos\omega t$ and the displacement (x) from mid-stroke of the piston is given by $x = r - y = r\cos\omega t$, which is an equation for simple harmonic motion—the instantaneous acceleration being $\ddot{x} = -r\omega^2\cos\omega t = -\omega^2 x$, i.e. the acceleration towards the mid-position is proportional to the displacement from it. Whereas if $R = r/l$ is *not* small enough the neglected but only the first

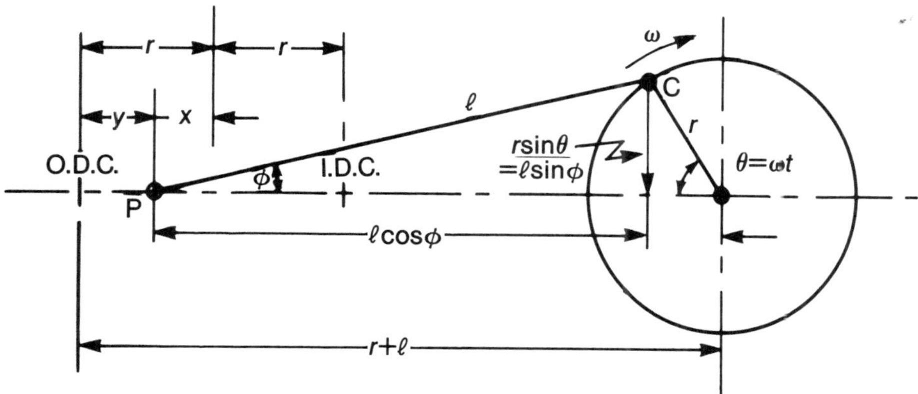

Fig. 1.3(d) Connecting rod and crank or slider-crank linkage

powers of R are included, then

$$y = r(1 + \tfrac{1}{4}R) - r\cos\omega t - r(\tfrac{1}{4}R)\cos 2\,\omega t$$

and $x = r - y = r\left\{\cos\omega t + \dfrac{R}{4}(\cos 2\omega t - 1)\right\}$ with instantaneous acceleration $\ddot{x} = -\,r\omega^2(\cos\omega t + R\cos 2\omega t)$ which is not that $(-\,\omega^2 x)$ of simple harmonic motion.

Thus, the force required to accelerate the reciprocating parts of mass m is $F = m\ddot{x} = mr\omega^2\cos\omega t + mr\omega^2 R\cos 2\omega t =$ (primary + secondary) force. The primary force has a frequency equal to that of the crank, and the secondary force a frequency twice that of the crank.

(b) If $r = 40$ mm and $l = 100$ mm for an engine whose piston has a mass of 1 kg, the maximum acceleration and retardation of the piston occurs at the ends of the stroke, and when the crank rotates at 3000 rev/min is,

$$a = r\omega^2(1 + R) = 40\text{ mm} \times \left(50\frac{\text{rev}}{s}\right)^2(1 + 0.4)$$

$$= 1.4 \times 10^5\text{ mm}\frac{\text{rev}^2}{\text{s}^2}\left[\frac{\text{m}}{10^3\text{ mm}}\right]\left[\frac{4\pi^2}{\text{rev}^2}\right] = 5.53\text{ km/s}^2,$$

and the maximum inertia force on the piston is given by

$$F = ma = 1\text{ kg} \times 5.53\frac{\text{km}}{\text{s}^2}\left[\frac{\text{N s}^2}{\text{kg m}}\right] = 5.53\text{ kN as against}$$

$$1\text{ kg} \times 0.04\text{ m}\left(\frac{50\text{ rev}}{\text{s}}\right)^2\left[\frac{4\pi^2}{\text{rev}^2}\right] = 3.95\text{ kN for s.h.m.}$$

1.4 RELEVANT LAWS OF MECHANICS

Any detailed study of mechanical vibrations requires a knowledge of certain fundamental principles of dynamics. Hence, the following summary or recapitulation (also see Appendix A).

The three laws of motion of Newton are laws of kinetics concerning the forces causing the motion of bodies. They can be appreciated in connection with 'inertia' which implies 'reluctance to be accelerated'. Thus, inertia is that property or attribute of matter by which it continues in its state of rest or of uniform motion in a straight line (the first law) unless that state is changed by external force—the force being proportional to rate of change of momentum (the second law which links the concepts of force, acceleration and mass); and the third law may be stated simply as 'action and reaction are equal and opposite'.

These laws apply to angular as well as to linear motion—torque being analogous to force, angular velocity and angular momentum to linear velocity and linear momentum, respectively, all of which are vector quantities, i.e.

have direction as well as magnitude (also see Appendix B1). Thus, because the force exerted by a mass on a spring is equal and opposite to the force exerted by the spring on the mass (and similarly with damping forces) it is usually desirable, for the sake of clarity, to isolate a mass diagrammatically and show the instantaneous forces acting upon it (i.e. *in free body diagrams*)—see Figs. 0.5(b), 1.3(c), 5.1(a) etc) before applying the second law in the form: force = mass × acceleration, to formulate the equation of motion (e.g. see Figs. 2.2(a), 3.2(a), 4.1(a), 5.5(c)) which involves 'absolute' acceleration. It is necessary to be quite clear as to whether motion is relative to a moving support (see Section 5.3B and Fig. 5.3(d)) or to the earth (i.e. 'absolute')—the earth being regarded as an infinite mass in a system or as a fixed support to which other masses are attached by elastic material.

The three laws of conservation fundamental to vibrational systems (which can be defined arbitrarily by enclosure within convenient boundaries—real or imaginary) concern matter, momentum and energy, namely.

(I) conservation of matter, i.e. matter is not destroyed (m is constant, or $\delta m = 0$); thus, there is always a mass balance;

(II) conservation of momentum, i.e. there is a momentum-impulse balance ($\delta(mv) = F\delta t$ and $\delta(I\omega) = T\delta t$ are consequences of Newton's 2nd law); thus, vectorially, momentum change is equal to impulse (also see Appendix A1(b)),

(III) conservation of energy, e.g. energy change of an isolated system is equal to the sum of the heating and working ($\delta E = \delta Q - \delta W$ which is zero if no energy, δQ, is added by heat-transfer and no external work, δW, is done *by* the system); thus, there is always an energy balance.

There are, however, many ways of expressing these basic laws in symbols dependent on definitions and type of problem, which may involve solids, liquids or gases—e.g. if the system is a body of constant mass supported by a spring and freely vibrating without dissipation or addition of energy, the sum of the kinetic energy (T) and the potential energy (V) of this 'conservative' system is constant, or $\frac{\mathrm{d}}{\mathrm{d}t}(T + V) = 0$ in accordance with law (III)-(see Example 1.4(vi).

EXAMPLE 1.4(i)—illustrative of the laws of conservation of momentum and energy

Estimate (a) the velocity of a bullet by firing it into a block of wood or sand bag suspended as a pendulum, (b) the energy dissipated and (c) these magnitudes in a particular case.

Referring to Fig. 1.4(a), the law of conservation of momentum yields

$$(m_1 v_1 + 0) = (m_1 + m_2)v_3$$

or

$$v_1 = \left(1 + \frac{m_2}{m_1}\right)v_3$$

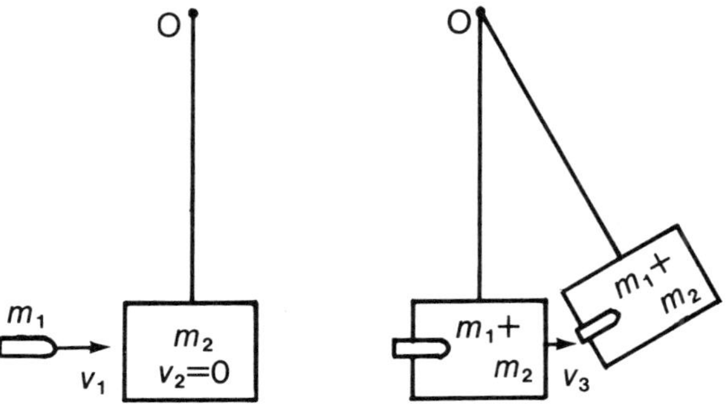

Fig. 1.4(a) Ballistic pendulum

and (b), the law of conservation of energy yields $\frac{1}{2}m_1 v_1^2 = \frac{1}{2}(m_1 + m_2)v_3^2$ + loss of energy i.e. the energy dissipated $= \frac{1}{2}\frac{m_2}{m_1}(m_1 + m_2)v_3^2 = \left(\frac{m_2}{m_1 + m_2}\right)\frac{1}{2}m_1 v_1^2$

(c) A bullet of mass 50 g fired into a sand bag of mass 50 kg suspended on a thin rod 1 m long caused the straight rod to deflect from its vertical position by a maximum angle of 14°6′. Neglecting the mass of the rod, kinetic energy of bullet + bag at lowest point = potential energy at highest point, i.e.

$$\tfrac{1}{2}(m_1 + m_2)v_3^2 = (m_1 + m_2)gh$$

where

$$h = (1 - \cos 14°6')\text{m} = (1 - 0.970)\text{m} = 30\,\text{mm}$$

Hence,

$$v_3 = \sqrt{(2gh)} = \sqrt{\left(60\,\text{mm} \times \frac{9807\,\text{mm}}{\text{s}^2}\right)} = 766\,\text{mm/s}.$$

Equating the momentum before and after impact gives

$$50\,\text{g}v_1 + 0 = (50 + 50 \times 10^3)g \times 0.776\ \text{m/s}$$

i.e. $v_1 = (1 + 10^3) \times 0.776\,\text{m/s} = 776\,\text{m/s}$ = the velocity of the bullet before striking the bag.

The loss of kinetic energy on impact is

$$\tfrac{1}{2}(m_1 v_1^2 - (m_1 + m_2)v_3^2) = \left(\frac{m_2}{m_1 + m_2}\right)\tfrac{1}{2}m_1 v_1^2 = \left(\frac{50 \times 10^3 \times 25}{50 + 50 \times 10^3}\right)g(776)^2\frac{\text{m}^2}{\text{s}^2}$$

$$= \left(\frac{25}{1 + 10^3}\right)(776)^2\,\text{kg}\frac{\text{m}}{\text{s}^2} = 15039\,\text{Nm} = 15.04\,\text{kJ}.$$

Particle dynamics are based on Newton's 2nd law of motion ($F = ma$); and a 'rigid' body may be regarded as being made up of a number of pairs of particles under the action of external and internal forces—the internal forces being equal and opposite in pairs so that the internal (including strain) energy remains constant, and only the translational and rotational energy of the body as a whole change (also, see Appendix A.3). An element of mass is treated as a particle when writing down the differential equations of linear motion—'mass' being a generic term which, in particular, is used to designate the coefficient by which the acceleration of a rigid body is multiplied to obtain the resultant force that accelerates the body in accordance with Newton's second law of motion ($F = ma$) and, therefore, has dimensions $[m] = [\mathrm{M}] = [\mathrm{FL}^{-1}\mathrm{T}^2]$. Similarly, for rotational motion ($T = I\alpha$) in which the displacement co-ordinate has the dimension of a plane angle whose unit is usually the dimensionless radian (see Section 0.3); and that of angular acceleration rad/s^2. The analogous 'rotational mass' ($I = T/\alpha$) has dimensions $[I] = [\mathrm{FLT}^2]$ or $[\mathrm{ML}^2]$, i.e. in dynamics, I represents the 'second moment of mass' but usually designated 'moment of inertia' in engineering nomenclature, whereas in statics (e.g. bending of beams $\dfrac{M}{I} = \dfrac{f}{y} = \dfrac{E}{R}$, see Appendix A5) I represents the second moment of area and, therefore, has dimensions $[\mathrm{L}^4]$.

External forces acting on a body fall into three categories, namely:

(i) gravitational force ($W = mg$) or weight,

(ii) forces caused by contact with other bodies (e.g. $f = kx$ exerted by a spring) and fluids (e.g. $f = cv = c\dot{x}$ of a viscous fluid), and exciting or applied forces (e.g. a varying or harmonic force $f = F\sin\omega t$ or $F\cos\omega t$ which commences when $t = 0$ and repeats after a period of time $\tau = 2\pi/\omega$).

(iii) forces causing (or inertia forces created by) rate of change of momentum $\left(f = \mathrm{d}(mv) = m\dfrac{\mathrm{d}v}{\mathrm{d}t} = ma \text{ for a constant mass}\right)$.

We shall see, in succeeding Sections, that, although many complex practical systems can be approximated for relatively straightforward analyses by setting up simpler or idealized systems having the common features of 'lumped' mass, elastic stiffness, viscous damping and one or two degrees of freedom, a more accurate determination of frequency of vibrations, say of lateral or transverse vibration of beams (see Section 6.1), requires that they be treated as continuous-mass or multi-mass systems.

EXAMPLE 1.4(ii)—illustrative of the four stages in engineering calculations

A mass m suspended on a light helical spring of stiffness k vibrates with a natural frequency of 1.6 Hz. With an extra 0.5 kg added to the mass, the cyclic frequency is reduced to 1.3 Hz. Calculate the mass m and the stiffness of the spring (see Fig. 2.2(a)).

(i) *Assumptions*: (a) the effect of the mass of the spring is negligible,
(b) the spring operates within its elastic limit,
(c) air resistance and inherent damping is negligible,
(d) the mass is constrained to a single degree of freedom.

(ii) *Scientific laws*: (a) Hooke's law applies to the spring.
(b) Newton's 2nd law of motion applies to the mass, followed by
(c) Equations of constraint.

(iii) *Mathematics*: $F = ma$ or $-kx = m\frac{d^2x}{dt^2}$, as deduced from the free-body diagram on the right of Fig. 2.2(a), where x is the displacement of the mass from the static-equilibrium position, and k is the spring stiffness (i.e. force per unit displacement).

Hence, $\frac{d^2x}{dt^2} + \omega^2 x = 0 = [D^2 + \omega^2]x$, where $\left(\frac{\omega}{\text{rad}}\right)^2 = k/m$, and the cyclic frequency $f = \frac{1}{2\pi}\sqrt{\left(\frac{k}{m}\right)}$—(see also, Example 0.5(b)).

(iv) *Arithmetic*: In the first case, the frequency

$$f_1 = \frac{1}{2\pi}\sqrt{\left(\frac{k}{m}\right)}\text{cycle} = 1.6 \text{ cycle/s} = 1.6 \text{ Hz},$$

and, in the second case, $f_2 = \frac{1}{2\pi}\sqrt{\left(\frac{k}{m + 0.5\text{ kg}}\right)}$ cycle = 1.3 cycle/s = 1.3 Hz

Eliminating k, we get

$$\frac{m + 0.5\text{ kg}}{m} = \left(\frac{f_1}{f_2}\right)^2 = \left(\frac{1.6}{1.3}\right)^2 = 1.513.$$

Hence,

$$m = \frac{0.5}{0.513}\text{kg} = 0.975 \text{ kg}$$

and

$$k = m\omega_1^2 = 0.975 \text{ kg}\left(\frac{3.2\pi}{s}\right)^2 \left[\frac{\text{N s}^2}{\text{kg m}}\right] = 98.5 \text{ N/m or } 0.985 \text{ N/cm}.$$

The predictions of theory can, of course, easily be verified (or otherwise) in the laboratory by measurements and observations on the actual system (See Section 6.4 for the six stages in laboratory work).

EXAMPLE 1.4(iii)—Clarification of the term 'inertia force' and its synonyms

In dynamics, 'inertia' may be thought of as 'reluctance to be accelerated'. Newton's second law of motion, $F = ma$ for a constant mass, may be written $F + (-ma) = 0$. Hence, since the quantity, $-ma$, appears in this equation of dynamic equilibrium, and having dimensions MLT^{-2}, it may be treated as a force. Thus, $-ma$ is called (perhaps unfortunately) by different names, e.g. inertia force, reversed-effective force, dynamic force, the d'Alembert force, as a concept whereby certain dynamical systems can be 'replaced' by static equivalents loaded by so-called 'reversed-effective' forces. A simple case is that of the conical pendulum of Fig. 1.4(b) in which the free-body diagram on the right-shows that the centripetal force (F_H) required to maintain a mass m rotating at

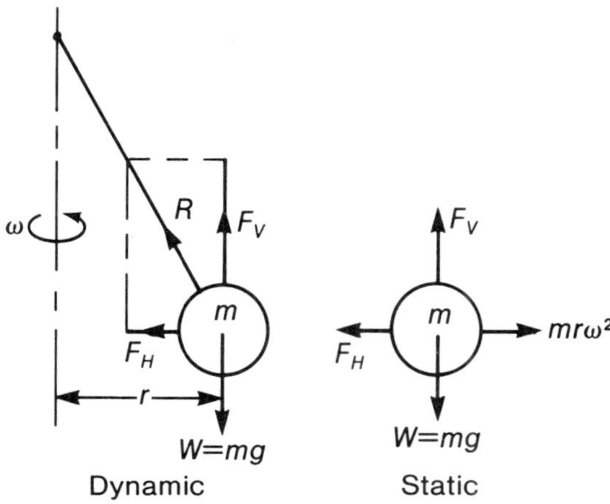

Fig. 1.4(b) Conical pendulum

radius r with an angular velocity, ω is $mr\omega^2$ (see Appendix A1(a)). Thus, $F_F - mr\omega^2 = 0$ which is equivalent in statics to a loading, inertia force or reversed-effective force, of $mr\omega^2$ as shown in Fig. 1.4(b).

It is, however, better to distinguish between dynamics and statics, and to treat them as such without searching for 'equivalents'.

Three derivatives concerning rates of change

Useful particular consequences of laws (II and III) in mechanics are.

(1) The rate of change of linear momentum of a rigid body in any direction is equal to the sum of the external forces in the same direction $\left(\text{i.e.} \frac{\mathrm{d}}{\mathrm{d}t}(mv) = \sum F = ma\right.$ for a body of mass m moving with linear acceleration a in the direction of $\sum F$. Also, see Appendix A.1(b)$\Big)$.

(2) The rate of change of angular momentum of a body about any fixed axis of rotation (o) is equal to the moment of the external forces on the body about that point i.e the equation relating magnitudes is $T_o = I_o\alpha = I_o\ddot{\theta}$ if the body is constrained to swing about a fixed axis through point o (see Example 1.4(iv)); and $T_G = I_G\ddot{\theta}$ for a body (see Appendix A2 and 3) moving freely in two dimensions where the reference point, G, is the centre of mass or centre of gravity of the body as in Example 1.4(vi).

(3) The rate of increase of total kinetic energy (translational and rotational) of a rigid body is equal to the rate at which external forces do work on the body if no energy is dissipated by friction. The equation of magnitudes (numbers × units—see Appendix B1) maybe expressed as $T_o\omega = (I_o\ddot{\theta})\dot{\theta}$ if the body is constrained to swing about a fixed axis through point o; and, in the case of a rigid body of mass m moving freely in two dimensions, as

$T_G\omega + Fv_G = (I_G\ddot{\theta})\dot{\theta} + (ma_G)v_G$, where F is the resultant component force acting at the centre of mass, G, in the direction of its linear velocity v_G and linear acceleration a_G (see Appendix A.2 and 3).

The action of forces on a body of elastic material can not only change the motion of the body as a whole but also cause changes in strain energy stored in the body. Should any part of a physical or mechanical system be disturbed from its static-equilibrium position, the resulting motion can be predicted if the solution to the appropriate differential equation of motion can be found (see Appendix B11). Such an equation is set up by equating the sum of the instantaneous values of the inertia, damping and elastic forces to the exciting or externally-applied force. As regards translational and rotational motion, it will become apparent (see Example 1.4(iv) & Section 2.2) that, as far as the mathematics is concerned, there is no difference algebraically between forces and translational displacements on the one hand and torques and angles of twist, respectively, on the other.

Since, in this book, we confine our attention to systems in which

(a) mass, m, and moment of inertia, I, are constant,

(b) restoring forces, kx or $k\theta$ (see Figs. 2.2(a) and (b)), are caused by the elastic properties of elements in the system within their elastic limits or by gravity, e.g. stiffness, k, is constant in springs and torsion bars; and in pendulums the restoring force is due to gravity,

(c) damping coefficient, c, is constant, e.g. damping force, being proportional to velocity in viscous damping, is cv or $c\dot{x}$ (see Examples 3.3(i) and (ii)),

we shall, therefore, be dealing with linear first $\left(\frac{dx}{dt}\right)$ and second $\left(\frac{d^2x}{dt^2}\right)$ order differential equations with constant coefficients (m, k, c), and with graphical or vector representations of such equations—'linear' meaning linear in the independent variable, x, and its derivatives $\left(\frac{dx}{dt}\right)$ or $\dot{x}$, $\left(\frac{d^2x}{dt^2}\right)$ or $\ddot{x}$, i.e. they appear only in the first degree and are not multiplied together (see Appendix B11).

We may also note that the total kinetic energy of a rigid body moving freely in two dimensions is the sum $(\frac{1}{2}mv_G^2 + \frac{1}{2}I_G\omega^2)$ of the translational and rotational kinetic energies—as proved in Appendix A.3.

EXAMPLE 1.4(iv)—illustrative of angular s.h.m. and conservation of energy

(a) Deduce an expression for the periodic time of a rigid body making small oscillations about a fixed horizontal axis, neglecting friction. Also, show that the sum of the kinetic and potential energies remains constant.

(b) A concentric spoked-wheel weighing 278 N pivoted on a knife edge at the inner rim 150 mm from the geometric centre makes small oscillations at the rate of 4/min. Estimate the moment of inertia of the wheel about its geometric axis (i.e. about its centre of gravity).

(a) Referring to Fig. 1.4(c), which is virtually a compound pendulum—the mass

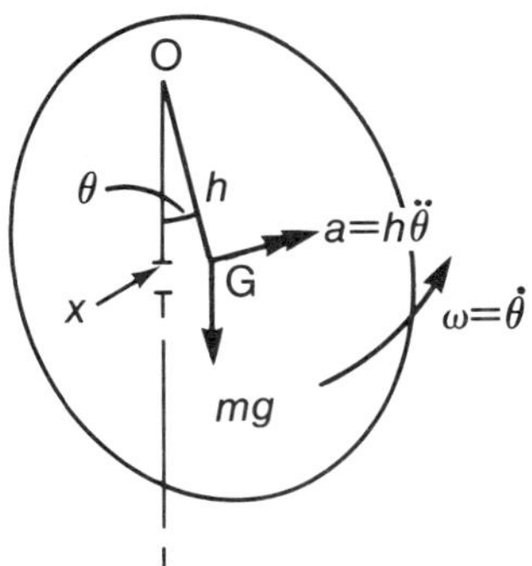

Fig. 1.4(c) Compound pendulum

being unevenly distributed along its length, first assume the axis through o is frictionless, and applying derivative (2), we deduce for angular motion that $T_o = I_o\ddot{\theta} = -mgh\sin\theta$ or, using operator D, we have $[D^2 + \omega_n^2]\theta = \left[D^2 + \frac{gh}{k_o^2}\right]\theta = 0$ for small oscillations. The solution, $\theta = C\cos(\omega_n t - \beta)$, as shown in Appendix B12, yields a periodic time $\tau = 2\pi/\omega_n = 2\pi\sqrt{\left(\frac{k_o^2}{gh}\right)}$, where k_o is the radius of gyration, and $k_o^2 = k_G^2 + h^2$.

We may note that the equation from derivative (2) can be re-written $I_o\dot{\theta}\ddot{\theta} = -mgh\,\dot{\theta}\sin\theta$ which, when integrated, becomes $\frac{1}{2}I_o\omega^2 = mgh\cos\theta = mg(h - x)$, i.e. $\frac{1}{2}I_o\omega^2 + mgx = mgh$ = constant (which is in accordance with law (III) and derivative (3) in this Section), i.e. kinetic + potential energy = constant, as expected of this conservative system.

(b) Taking the result of frictionless small oscillations, we may write, $k_o = \frac{\tau}{2\pi}\sqrt{(gh)}$ i.e.

$$k_o = \frac{60s}{48 \times 2\pi}\sqrt{\left(9807 \times 150\frac{\text{mm}^2}{\text{s}^2}\right)} = 241 \text{ mm.}$$

Also, since $I_G = I_o - mh^2$, or $k_G^2 = k_o^2 - h^2 = 391 \times 91 \text{ mm}^2$, then

$$I_G = \frac{W}{g}k_G^2 = \frac{278 \text{ Ns}^2}{9807 \text{ mm}} \times 391 \times 91 \text{ mm}^2 = 1 \text{ Nms} \quad \text{or} \quad 1 \text{ kg m}^2.$$

EXAMPLE 1.4(v)—illustrative of how *I* of complex shapes can be found experimentally
Deduce a formula to give the radius of gyration and moment of inertia about the centre of mass of bodies of complex shape when suspended by wires and set to perform small angular oscillations.

Except for simple shapes, moments of inertia are never calculated but are estimated from observations of the dynamic response to angular acceleration—e.g. the frequency of the body when oscillating freely as a filar pendulum with small amplitude. This is an important application of pendulums for the determination of I_G of marine propellers, turbine rotors, gear wheels, etc. which can be suspended by wires equidistant from the

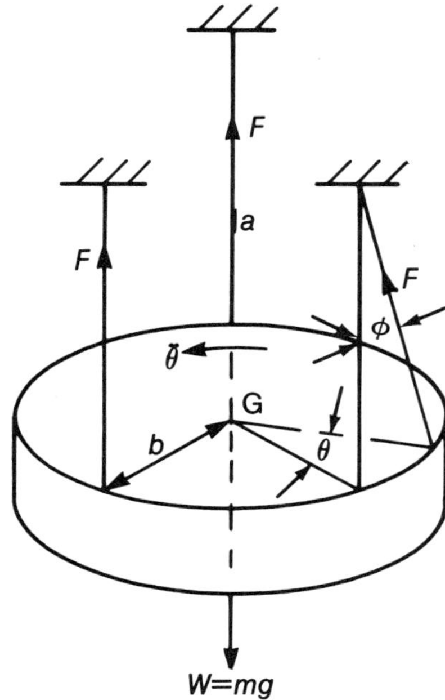

Fig. 1.4(d) Trifilar suspension

centre of gravity, and observing the periodic time for small oscillations. Generally, three wires give the best stability, but the formula will be found to be independent of the number of wires.

In the case of the trifilar suspension of Fig. 1.4(d) the rim displacement for small angles of ϕ is $\theta b = \phi a$ and the force in each wire is $F = mg/3$. Also, the restoring force tangential to the rim is $3F' = 3F \sin\phi = mg\phi$ for small angles.

Hence, applying $T_G = I_G\ddot{\theta}$, we have $mk_G\ddot{\theta} = -3F'b = -mg\dfrac{b^2}{a}\theta$ for small amplitudes,

i.e. $\ddot{\theta} + \left(\dfrac{b^2}{a}\dfrac{g}{k_G^2}\right)\theta = 0 = \ddot{\theta} + \omega_n^2\theta$, where ω_n is the natural frequency of the undamped oscillations. The periodic time $\tau = 2\pi/\omega_n$ (see Section 0.4) can be estimated by averaging the time of, say, 10 complete cycles, and hence, the radius of gyration of the body about G can be found from $k_G = \tau\dfrac{b}{2\pi}\sqrt{(g/a)}$ and $I_G = \dfrac{Wb^2}{a}\left(\dfrac{\tau}{2\pi}\right)^2$.

EXAMPLE 1.4(vi)—illustrative of the use of the energy method for finding frequency and periodic time

A cylinder of radius r and weight W rolls without slipping on a cylindrical surface of radius R. Derive a formula for estimating the periodic time for small oscillations about the lowest point, neglecting friction.

The method of using an energy balance (law III or derivative 3 previously stated) is relatively easy of application for this and many other problems concerning oscillations and vibrations. From the free-body, roll-oscillating, cylinder of Fig. 1.4(e) we deduce

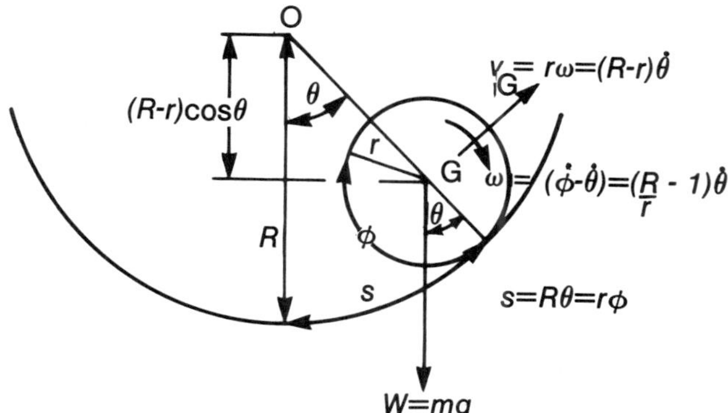

Fig. 1.4(e) Cylinder, roll-oscillating

that, since the restoring force is due to gravity, the 'potential' energy of the cylinder, depending on height above the lowest point, is $V = W(R-r)(1-\cos\theta)$ and the kinetic energy (see Appendix A3(c)) of the cylinder is

$$T = \tfrac{1}{2}mv_G^2 + \tfrac{1}{2}I_G\omega^2 = \frac{1}{2}\left\{m(r\omega)^2 + \left(\frac{mr^2}{2}\right)\omega^2\right\}$$

$$= \tfrac{3}{4}mr^2\omega^2 \quad \text{or} \quad \frac{3}{4}\frac{W}{g}(R-r)^2\dot{\theta}^2$$

Hence, by the law of conservation of energy, $T + V =$ constant for the conservative system which, by definition, implies no dissipation of energy by friction. Hence,

$$\frac{d}{dt}(T+V) = 0 = \frac{3}{2}\frac{W}{g}(R-r)^2\dot{\theta}\ddot{\theta} + W(R-r)\dot{\theta}\sin\theta$$

or

$$\ddot{\theta} + \frac{2g\sin\theta}{3(R-r)} = 0.$$

For small oscillations $\sin\theta \approx \theta$ and hence, using operator D^2 for $\frac{d^2}{dt^2}$, we get the second order linear differential equation $\left[D^2 + \frac{2g}{3(R-r)}\right]\theta = 0$ or $[D^2 + \omega_n^2]\ \theta = 0$, where $\omega_n^2 = \frac{2g}{3(R-r)}$ and a solution $\theta = C\cos(\omega_n t - \beta)$ from which we deduce (see Appendix B12(i)) that the periodic time is

$$\tau = \frac{2\pi}{\omega_n} = 2\pi\sqrt{\left(\frac{3(R-r)}{2g}\right)}.$$

1.5 HARMONIC FORCES AND MOTION

A visual aid in the form of a vector and wave diagram for considering s.h.m. is shown in Fig. 1.5(a). The vector X (of constant length but varying in direction) rotating at constant angular velocity or angular frequency, ω, gives simple harmonic motion to the point of projection, $x = X \sin \omega t$, on the vertical diameter. Thus, any harmonically-varying quantity, e.g. $y = Y \sin(\omega t + \psi)$, can be represented by a rotating vector as can a harmonic force $f = F \cos(\omega t + \beta)$.

It is clear that a circular function (sine or cosine) repeats itself in 2π radians, and a cycle is completed when $\omega\tau = 2\pi$, where τ is the periodic time and ω the radian rate. The instantaneous velocity of the projection of rotating vector X is $\dfrac{dx}{dt} = \dot{x} = \omega X \cos \omega t = xX \sin\left(\omega t + \dfrac{\pi}{2}\right)$, and its instantaneous acceleration is $\dfrac{d^2x}{dt^2} = \ddot{x} = -\omega^2 X \sin \omega t = \omega^2 X \sin(\omega t + \pi) = -\omega^2 x$, both of which are harmonic and, hence, can be represented by vectors rotating with the same angular velocity or frequency (ω in radians per unit of time) as the maximum displacement vector X, but in phase relationships $\dfrac{\pi}{2}$rad or 90° and π rad or 180°, respectively, ahead of it. It is useful to remember this sequence in time, namely, that acceleration preceeds velocity which, in turn, preceds displacement as shown in Fig. 1.5(a) in which the angles between the rotating vectors, or phase angles, are clearly seen on the abscissa.

The instantaneous acceleration $a = \ddot{x} = -\omega^2 x$ can be written $\dfrac{d^2x}{dt^2} + \omega^2 x = 0$ or as $[D^2 + \omega^2]\, x = 0$, where D is an 'operator' indicating that the operation of differentiation with respect to time has to be performed, and D^2 indicates double differentiation (see Appendix B10).

Thus, sine and cosine terms indicate harmonics which can be inter-

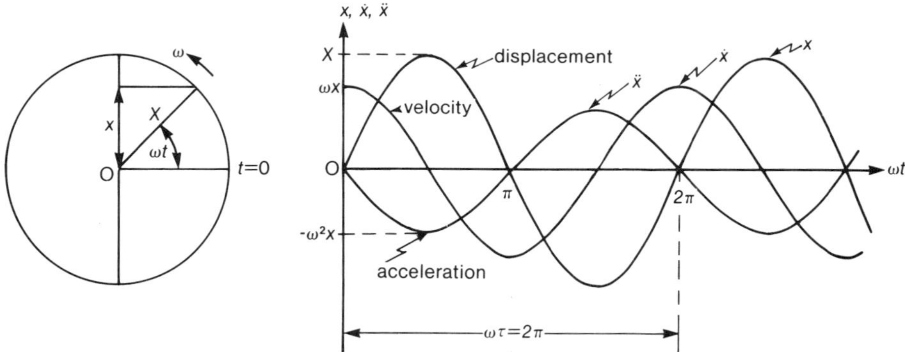

Fig. 1.5(a) Diagrammatic of s.h.m.

changed allowing for an angular phase difference of $\pi/2$ radians,—e.g. $\cos \omega t = \sin(\omega t + \pi/2)$. Generally, the term 'simple harmonic' is applied to motion, although a force can be referred to as acting harmonically,—e.g. a centripetal force which acts radially inwards with constant magnitude ($F = mr\omega^2$) but varying direction, and has a component, f, which acts harmonically along one particular line, e.g. $f = F \sin \omega t$.

From Fig. 1.5(a) it will be clear that if the angular frequencies of two motions (or forces) acting on a particular point are different, the vectors representing the two will, in general, be separated by a phase angle, ϕ, at any instant. Therefore, unless the vectors rotate at the same speed, i.e. have the same angular frequency, the phase angle ϕ will vary between 0 and 2π radians. At one instant the vectors will coincide and give a maximum resultant amplitude of motion, whereas at another instant they will be diametrically opposed giving a minimum resultant amplitude. The 'beat' frequency will be the difference between the frequencies of the two component vectors (as in Example 1.5(ii)). Thus, beats are a manifestation of coupled vibrations or oscillations when the two are nearly the same frequency resulting in a waxing and waning of the sum or resultant.

Alternative to the vector and wave diagrams of Fig. 1.5(a) we may note that a rotating vector can be represented mathematically (see Appendix B9) by a complex quantity involving circular or exponential functions. For example, using j to represent $\sqrt{-1}$, a vector of length r rotating with angular velocity ω can be represented as

$$r\varepsilon^{j\omega t} = r(\cos \omega t + j \sin \omega t) = x + jy, \text{ where } \omega t = \tan^{-1}(y/x)$$

and $r = \sqrt{(x^2 + y^2)}$ as shown in Fig. 1.5(b). The real and imaginary parts at any instant are obtained from the perpendicular projections of vector r on to the horizontal and vertical axes, namely, $r \cos \omega t$ and $r \sin \omega t$, respectively. Both $(\cos \omega t + j \sin \omega t)$ and $\varepsilon^{j\omega t}$ may be regarded as "operators" for turning vector r by an angle ωt.

Thus, using the complex exponential functions $x = X\varepsilon^{j\omega t}$, we deduce that $\dot{x} = j\omega X\varepsilon^{j\omega t}$

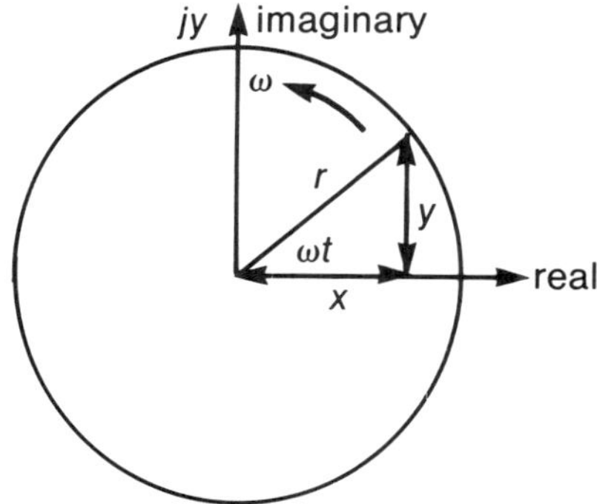

Fig. 1.5(b) Complex quantity $r\varepsilon^{j\omega t} = x + jy$

and the differential equation for s.h.m., $\ddot{x} = -\omega^2 X\varepsilon^{j\omega t} = -\omega^2 x$, i.e. $\frac{d^2x}{dt^2} + \omega^2 x = 0 = [D^2 + \omega^2]x$ which shows that $x = X\varepsilon^{j\omega t}$ is a harmonic function, and that both the real ($X\cos\omega t$) and imaginary ($X\sin\omega t$) components satisfy the differential equation separately. Hence, if a harmonic motion is represented by either of the circular functions, $x = X\sin\omega t$ or $X\cos\omega t$, we can (if more convenient) replace it by the exponential form, $\varepsilon^{j\omega t}$ knowing that either component (real or imaginery) will independently satisfy the differential equation for s.h.m., the general solution of which is

$$x = A\sin\omega t + B\cos\omega t = C\sin(\omega t + \phi) = C\cos(\omega t - \psi)$$

or, alternatively, $x = \alpha\varepsilon^{j\omega t} + \beta\varepsilon^{-j\omega t}$ (See Appendix B6 and B12).

We may conclude, therefore, that, although it is often preferable, easier and more familiar to use trigonometrical rather than complex exponential forms, $x = X\varepsilon^{j\omega t}$ offers a convenient mathematical way of representing a rotating vector of constant length X and that, since $[D^2 + \omega^2]x = 0$, the projection of the tip of the vector on to the x axis moves with s.h.m. because $\ddot{x} \propto -x$. Similarly, the harmonic force $F\sin\omega t$ corresponds to the imaginary part of the complex exponential $F\varepsilon^{j\omega t}$ and $F\cos\omega t$ to the real part; and likewise for the harmonic torque $T\varepsilon^{j\omega t}$.

However, to proceed with vibration analysis via mathematics only would be rather formidable and less revealing compared with the complementary method of vector representation on phase diagrams. The latter also facilitate appreciation of the physical aspects of dynamic systems—particularly those of damped and forced vibrations, as will be dealt with in Chapters 3 and 4. The differential equation of motion is, however, fundamental and inevitable whether one proceeds entirely mathematically or with the help of vector phase diagrams.

We may also note that, without the introduction of new concepts, vibration analysis can be applied in many important fields of engineering other than that of seeking to suppress mechanical vibration, e.g. on the mass-inductance analogy (also, see Appendix C9) the electrical quantities of inductance, resistance, capacitance, current and electromotive force are analogous to mass, damping coefficient, reciprocal of stiffness, velocity and applied force, respectively. Likewise the equations relating to responses and stability in servomechanisms and simple control systems can be discussed in terms of the basic concepts of vibration analysis because servomechanisms may contain standard mechanical links such as shafts, cranks, gears and cams as well as electrical and hydraulic linkages in which tubing (analogous to electrical cables) carry fluids under pressure to operate rams or motors—a fluid pump being analogous to an electrical generator, and valves to electrical switches, etc. If a signal is initially mechanical and strong it may directly operate the valve of a hydraulic system, but if it is weak there can be conversion into an electrical signal and amplification for control applications.

EXAMPLE 1.5(i)—illustrative of non-harmonic motion

Set up the differential equation for the large-amplitude non-harmonic horizontal oscillation of a mass m attached to the middle of a vertical taut wire of length $2y$, cross-sectional area a, initial tensile force T_1 and modulus E but of negligble mass compared with m. Also show that the motion of the mass approaches simple harmonic for very small displacements.

Referring to Fig. 1.5(c) and assuming T_1 to be the initial tensile force in the wire, we deduce:

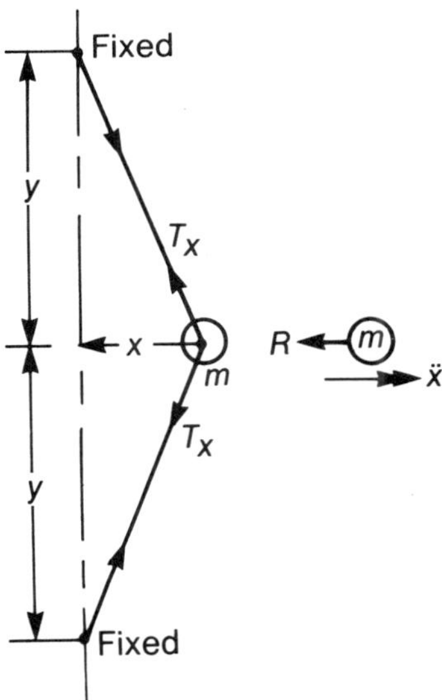

Fig. 1.5(c) Non-harmonic motion

(i) Additional strain on the wire owing to displacement x is $\dfrac{\sqrt{(y^2+x^2)}-y}{y} = e_x$ or

$$e_x = \frac{(y^2+x^2)-y^2}{y\{\sqrt{(y^2+x^2)}-y\}} = \frac{x^2}{2y^2}\begin{pmatrix}\text{neglecting } x^2\\ \text{compared with } y^2\end{pmatrix}.$$

(ii) Additional tensile force to produce this strain is $aEe_x = \dfrac{aEx^2}{2y^2}$.

(iii) The total tensile force in the wire when the displacement of the mass is x, is

$$T_x = T_1 + \frac{aEx^2}{2y^2}.$$

(iv) The restoring force, R, on the mass m is $R = 2T_x\left(\dfrac{x}{\sqrt{(y^2+x^2)}}\right) = \dfrac{2x}{y}T_x$, again

neglecting x^2 compared with y^2, then $R = \dfrac{2x}{y}T_1 + \dfrac{aEx^2}{y^3}$.

(v) Since (from the free-body diagram), $m\ddot{x} = -R$, $\dfrac{d^2x}{dt^2} + \left(\dfrac{2T_1}{my}\right)x + \left(\dfrac{aE}{my^3}\right)x^3 = 0$ is the differential equation for the non-harmonic motion of the mass.

In the case of very small displacements of the mass, i.e. when the initial tensile force, T_1, is sufficiently large, the last term can be neglected; and the vibration of the mass in a horizontal direction will be very nearly simple harmonic obeying $[\mathrm{D}^2+\omega_n^2]x=0$, $x = X\cos(\omega_n t+\beta)$ with a periodic time $\tau_n = \frac{2\pi}{\omega_n} = 2\pi\sqrt{\left(\frac{my}{2\pi T_1}\right)}$.

EXAMPLE 1.5(ii)—illustrative of the notion of the phase plane

Plot velocity against displacement for s.h.m.—i.e. plot 'phase plane' trajectories for simple harmonic motion.

It is usually simpler to deal with first order differential equations than any other—especially when numerical methods are used in solving non-linear differential equations. In the case of the second order linear differential equation for s.h.m., namely, $\frac{\mathrm{d}^2x}{\mathrm{d}t^2}+\omega^2x=0$, a mathematical ploy enables us to replace this by a differential equation of the first order by writing $\frac{\mathrm{d}x}{\mathrm{d}t}=y=\dot{x}$. Thus, $\ddot{x}=\dot{y}=\frac{\mathrm{d}y}{\mathrm{d}t}=-\omega^2x$ i.e. $\frac{\mathrm{d}y}{\mathrm{d}t}=\frac{\mathrm{d}y}{\mathrm{d}x}\frac{\mathrm{d}x}{\mathrm{d}t}$ $=y\frac{\mathrm{d}y}{\mathrm{d}x}=-\omega^2x$, where $y=\dot{x}=\frac{\mathrm{d}x}{\mathrm{d}t}$. Hence, $\omega^2x\,\mathrm{d}x+y\,\mathrm{d}y=0$ which is easy to integrate into $\omega^2x^2+y^2=b^2$ (a constant).

The plane in which x and y are Cartesian co-ordinates has been called the *phase plane*. Thus, the variation of the velocity, $\dot{x}=y$, with displacement, x, is described by a curve (an ellipse for s.h.m.) or trajectory in the phase plane, and the differential equation of the trajectory is $\frac{\mathrm{d}y}{\mathrm{d}x}=\frac{\dot{y}}{\dot{x}}=\frac{\dot{y}}{y}=-\frac{\omega^2x}{y}$. Fig. 1.5(d) shows that trajectory and, although the phase plane plot does not reveal the precise variation of x with t, we can deduce from it that the motion is periodic because on going round an ellipse we periodically return to the starting values of x and y. Also, the maximum amplitude (when $y=\dot{x}=0$) is b/ω.

Since $\frac{\mathrm{d}x}{\mathrm{d}t}=\dot{x}=y$, then $\mathrm{d}t=\frac{\mathrm{d}x}{y}$ and the period (taken round the ellipse) of the s.h.m. is

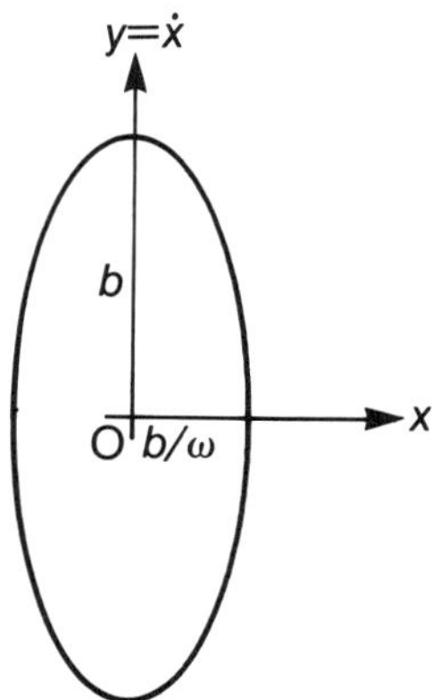

Fig. 1.5(d) Phaseplane for s.h.m.

$$\tau = 4\int_0^{b/\omega} \frac{dx}{\sqrt{(b^2 - \omega^2 x^2)}} = 2\pi/\omega \begin{pmatrix} \text{the integral being achieved via} \\ \text{the substitution } x = (b/\omega)\sin\theta \end{pmatrix}$$

In general, when a 'phase plane trajectory' is a closed curve, the corresponding motion is periodic but, as stated in Section 1.3, a periodic motion is not necessarily harmonic.

An alternative phase-plane representation of a single degree of freedom system executing vibration in accordance with the equation of motion $\ddot{x} + \omega^2 x = 0$, is a circle. For, the solution $x = X\sin(\omega t + \alpha)$, where the constants X (the amplitude) and α (the $t = 0$ phase angle) are determined by the initial conditions, yields $\frac{\dot{x}}{\omega} = X\cos(\omega t + \alpha)$.

Hence, $x^2 + \left(\frac{\dot{x}}{\omega}\right)^2 = X^2$ which, with x and $(\dot{x}/\omega)$ as Cartesian co-ordinates has a circular trajectory of radius X as in Fig. 1.5(d). Initial conditions x_0 and $\dot{x}_0/\omega$ when $t = 0$ determine the starting position P_0 and P represents the position when the displacement X is x and the velocity $\dot{x}$ at any time t as P moves anticlockwise round the circle. This form of the phase plane is easier to plot and, also, both axes have the dimension of a length.

For suggested further reading and study, see Appendix D, (1), (2), (3) and (9).

EXERCISES ON CHAPTER 1

1. Add the two harmonics $x_1 = 4\cos(\omega t + 10°)$ and $x_2 = 6\sin(\omega t + 60°)$.
2. Determine the sum of the two vectors $5\varepsilon^{j\pi/2}$ and $8\varepsilon^{j\pi/6}$.
3. Measuring instruments indicate that a body is vibrating harmonically with a frequency of 8 Hz and a maximum acceleration of 8 m/s^2. Calculate the amplitude of vibration using Hz to mean cycle/s = 2π rad/s = 2π/s.
4. A harmonic motion is given by the equation $y = 4\sin(5t + \beta)$. Find the harmonic equations for its two components—one leading it by 30° and the other lagging it by 80°.
5. At a certain dock the rise and fall of the tide is 3.6 m. High water occurs at 13.00 h and low water at 19 h 30 min. Assuming the rise and fall of the tide is simple harmonic, calculate the instant when the water is 0.9 m below high-water mark, and estimate the rate in mm/min at which the water is falling.
6. A connecting rod weighing 21.6 N was allowed to make small oscillations about a knife edge supporting the rod at the gudgeon pin bush. The centre of gravity was 250 mm from the line of support. Estimate the moment of inertia of the connecting rod about its centre of gravity if it oscillates at the rate of 39 per minute.
7. A target on rough sea rises and falls with the waves assumed in s.h.m. with periodic time 8 s and double amplitude (or height from crest to trough) of 3 m. The range is 2.1 km from a fixed gun which fires a missile with a horizontal speed of 700 m/s (i.e. Mach 2). If the gun is fired with its sighting correct when the target is at the top of a wave, estimate by what vertical height the missile will miss the target.

2

Undamped natural vibrations

2.1 METHODS OF ANALYSIS

Natural vibration occurs when an elastic system is disturbed from its static-equilibrium position, and thereafter is free from impressed or exciting forces, excepting the constant force of gravity on the system. Such natural vibrations are maintained by elastic forces in the system which will vibrate at its natural frequency (ω_n). The amplitude (X) will, however, gradually diminish to zero because of inevitable slight damping and consequent dissipation of energy, e.g. caused by hysteresis in solid materials, since no material is perfectly elastic. In many cases, natural vibrations are sufficiently persistent for the motion to be treated as undamped in 'conservative' systems, i.e. from which no energy is dissipated. The effect of damping on natural frequency, and the rate of decay of amplitude are dealt with in detail in Chapter 3.

Three methods of analysis available are: (i) $F = ma$ and $T = I\alpha$ for translation and rotation, respectively, and which relate to magnitudes of physical quantities (also see Appendix B1), (ii) kinetic energy (T) + potential energy (V, which may be due to position above a datum and/or deformation storing strain energy) is constant, if there is no friction, i.e. $\delta(T + V) = 0$ for a conservative system (see Section 1.4(III)), (iii) since V is a maximum when $T = 0$, and vice versa, $(V)_{max} = (T)_{max}$ which usually yields ω_n, the angular frequency of undamped natural oscillations. This, of course, assumes that the relationship between the two maxima is known in terms of the configuration of the vibrating system. If, however, this is not precisely known, an approximate method (due to Rayleigh) assumes suitable shapes (e.g. the static-deflection curves) to be the amplitude contours or maximum displacement shapes of beams in lateral vibration in order to calculate T and V before applying the equation $(T)_{max} = (V)_{max}$ to yield, relatively quickly, an approximate calculation of the lowest natural frequency of vibrating beams. These three methods are used in worked Examples in this chapter.

2.2 NATURAL FREQUENCY OF UNDAMPED SYSTEMS

In the simplest of vibrating or oscillating systems the position of the masses are completely specified by the value of a single co-ordinate (x or θ) representative of systems with one degree of freedom.

With the sign convention (see Section 0.2) that the downward direction is positive for displacement (hence, necessarily, for velocity and acceleration as well) we now show, once and for all, that the static-equilibrium position is the most convenient datum. Assuming one degree of freedom (i.e. vertical linear motion) and the effect of the mass of the spring to be negligible compared with that of the oscillating body then, referring to Fig. 2.2(a) and applying the second law of motion, we deduce (from the free-body diagram on the right) that $mg - ky = m\ddot{y}$ and $mg = kd = \text{constant}$, or

$$\ddot{y} + \frac{k}{m}\left(y - \frac{mg}{k}\right) = 0 = \ddot{y} + \frac{k}{m}(y - d).$$

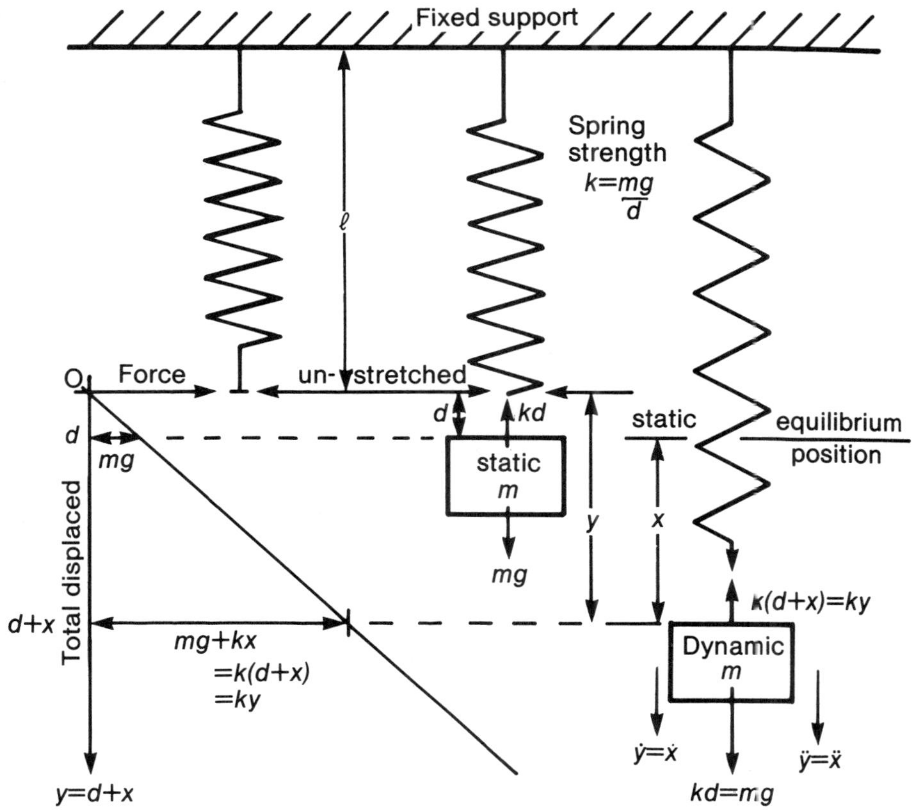

Fig. 2.2(a) Undamped natural vibration of a spring-mass system

Also, $x = y - d$ and $\ddot{x} = \ddot{y}$, since d is constant. Hence, relative to the static-equilibrium position (where the weight of the mass and force of the spring balance each other) we have $\ddot{x} + (k/m)\, x = 0 = [D^2 + \omega_n^2]x$, which is basic to simple harmonic motion as shown in Sections 0.5 and 1.5. Thus, *mg/k and d are eliminated if the static-equilibrium position is selected as the datum line for displacement*, x, about which line the oscillatory motion is symmetrical. *This fact is accepted without further mention in succeeding Sections.*

The solution of the differential equation (see Appendix B 12(i)) can be written in the trigonometrical form $x = X \cos(\omega_n t + \beta)$, where the undamped natural frequency $\omega_n = \sqrt{(k/m)} = \sqrt{(g/d)}$, and X and β are constants. Thus, the angular frequency; ω_n, and the cyclic frequency, $f_n = \omega_n/2\pi$, are independent of amplitude, X.

The general solution can be written in several forms (see Appendix B 9 and 12) three of which are

$$x = X\cos(\omega_n t + \beta) \text{ or } X \sin(\omega_n t + \alpha),$$

$$x = A \cos \omega_n t + B \sin \omega_n t \text{ and } x = C_1 \varepsilon^{j\omega t} + C_2 \varepsilon^{-j\omega t};$$

and are all harmonic functions of time. We may note that, although the solution must contain two arbitrary constants, they need not both be amplitudes (A and B) for, if more convenient, one can be an amplitude (X) and the other a phase angle (α or β).

Alternatively, from an energy point of view, the strain energy stored in the spring, when the mass is in the static-equilibrium position (also see Appendix A7) is $\frac{1}{2}mg\, d = \dfrac{1}{2k}(mg)^2$; and when the mass is a distance x below that position, the strain energy in the spring is $\dfrac{k}{2}(d + x)^2 = \dfrac{1}{2k}(mg + kx)^2$. Hence, the increase in strain energy in the system caused by an extension, x, of the spring from the static-equilibrium position is

$$\frac{1}{2k}\{(mg + kx)^2 - (mg)^2\} = \frac{1}{2k}\{2\, mgkx + k^2x^2\} = mgx + \tfrac{1}{2}kx^2.$$

The loss of energy by the system due to change of position (i.e. loss of height of the mass owing to extension x) is mgx. Hence, the net gain in 'potential' energy (strain + position) due to the extension x of the spring beyond the static-equilibrium position is $\frac{1}{2}kx^2$, which is the same as the strain-energy of an initially-unloaded spring deflected by an amount x. Thus, using the term 'potential' to include strain-energy and energy due to position, or height above a datum line, *we need only include the change in strain energy due to displacement from the static equilibrium position in calculating the potential energy (under quantity-symbol V) of a loaded elastic system*, since the effects of the gravitational field as regards energy due to position (height) have been shown to eliminate themselves.

During oscillation of the mass (Fig. 2.2(a)), an interchange of energy continually occurs—e.g. (i) when the mass passes through the equilibrium position, the kinetic energy (T) is a maximum and 'potential' (strain) energy (V) is zero, and (ii) when the mass has its maximum displacement (i.e. amplitude X) it is momentarily at rest and hence, the potential energy is a maximum whilst the kinetic energy is zero. Therefore, if damping is neglegible, $T_{\max} = V_{\max}$—i.e. $\frac{1}{2}m(X\omega_n)^2 = \frac{1}{2}kX^2$, and $\omega_n = \sqrt{(k/m)}$ showing that this energy method is shorter and simpler than the momentum ($\delta(mv) = F\delta t$ or $F = ma$—see Section 1.4II) method in this particular case of s.h.m.

Again, from another energy point of view (Fig. 2.2(a)) we deduce that the total potential energy of the mass at position x below the static-equilibrium position is $V_x = \frac{1}{2}k(x+d)^2 - mgx = \dfrac{k}{2}(x^2+d^2)$, and the kinetic energy of the mass at position x is $T_x = \frac{1}{2}m\dot{x}^2$.

Hence, for a conservative system (i.e. in free oscillation, neglecting frictional damping) $\dfrac{d}{dt}(T_x + V_x) = 0 = m\dot{x}\ddot{x} + kx\dot{x}$ or $m\ddot{x} + kx = 0$, which is the equation of motion obtained from this energy balance in accordance with the law of conservation of energy, and from which we again deduce that $\omega_n = \sqrt{(k/m)}$.

Torsional oscillation

In the case of undamped natural torsional oscillations caused by release from an initial angular displacement of, say, a cylindrical mass or disc of moment of inertia I secured to a cylindrical shaft of length L and second polar moment of area J and modulus of rigidity G as shown in Fig. 2.2(b), we may apply the formula relating magnitudes for torque. Thus, $T = I\alpha = I\ddot{\theta}$ and (see Appendix A8) since $\dfrac{T}{J} = \dfrac{G\theta}{L}$, $T = I\ddot{\theta} = -\dfrac{GJ\theta}{L}$ from the free-body diagram, or $I\ddot{\theta} + k\theta = 0$, where the torsional stiffness of the shaft is $k = \dfrac{T}{\theta} = \dfrac{GJ}{L}$ (also, see Appendix A6 and A8). Thus, $[D^2 + \omega_n^2]\theta = 0$, where $\omega_n^2 = \dfrac{k}{I} = \dfrac{GJ}{LI}$. The solution of the differential equation can be written as $\theta = A\cos(\omega_n t - \beta)$ in which A and β are constants (as in Appendix B12), and the cyclic frequency is $f_n = \dfrac{\omega_n}{2\pi} = \dfrac{1}{2\pi}\sqrt{\left(\dfrac{k}{I}\right)}$.

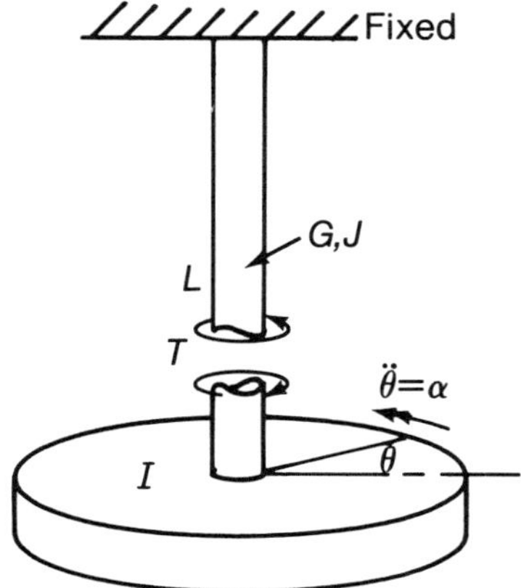

Fig. 2.2(b) Torsional oscillation

2.3 PHASE DIAGRAM FOR UNDAMPED NATURAL VIBRATIONS

Undamped natural or free vibrations are one category in four (categorised in Section 3.2), and are the easiest to deal with.

We have deduced from Fig. 2.2(a) that the instantaneous displacement x as a function of time t can be written in the trigonometrical form

$$x = X\cos(\omega_n t + \beta), \tag{1}$$

where X is the amplitude and β the initial phase angle, i.e. when $t = 0$. The corresponding equation for velocity of mass m is

$$v = \dot{x} = -\omega_n X \sin(\omega_n t + \beta) = V\cos(\omega_n t + \beta + \pi/2), \tag{2}$$

where V is the maximum value $(\omega_n X)$ of the velocity. Similarly, the equation for acceleration of mass m is

$$a = \ddot{x} = -\omega_n^2 X\cos(\omega_n t + \beta) = A\cos(\omega_n t + \beta + \pi), \tag{3}$$

where A is the maximum value $(\omega_n^2 X)$ of the acceleration.

These equations for x, v and a can be interpreted diagrammatically on the polar diagram of Fig. 2.3(a) which has a fixed vertical line OA drawn to represent zero time as on an ordinary clock, and an arrowed line (like a double-ended clock hand) imagined to be rotating about O at a constant angular rate ω_n. This clock-hand or 'time-line' (the only line imagined to rotate) starts at $t = 0$ and has an angular displacement $\omega_n t$ at any time t. The

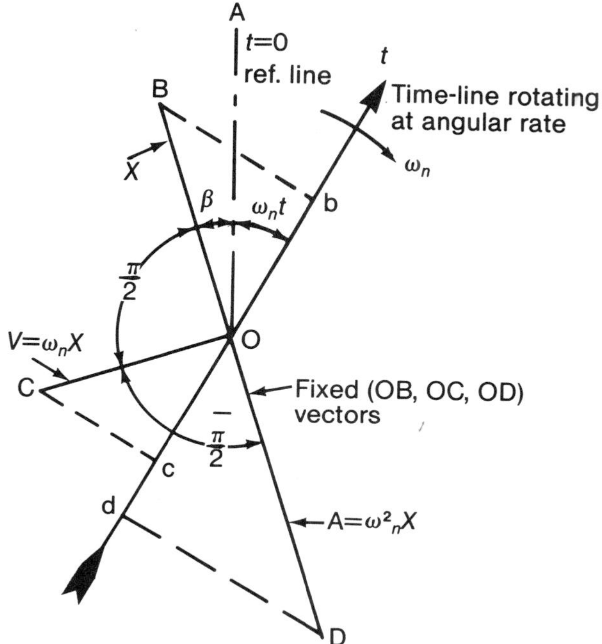

Fig. 2.3(a) Polar or phase diagram for undamped ($\zeta = 0$) natural oscillations

rotating time-line gives tangible meaning to the terms 'angular frequency (ω)' and 'phase rate'—both of which are used as abbreviations of the phrase 'frequency of rotation of the time line'.

Line OB is drawn at an angle β to the zero time-line (i.e. counterclockwise if β is positive so that when $t = 0$, $x = X \cos \beta$) and of length X, which is the maximum displacement of amplitude. Thus, the instantaneous displacement at time t, namely, $x = X \cos(\omega_n t + \beta)$, is given by length Ob which is *positive* (i.e. the displacement of m is downwards, or below the mean oscillatory position) because Ob falls *towards the head* of the time-line.

Line OC is drawn at $\pi/2$ rad, or 90°, counter-clockwise from OB and of length $\omega_n X = V$, which is the maximum value of the velocity of the mass m. The instantaneous velocity at time t, namely $v = V \cos(\omega t + \beta + \pi/2)$, is given by length Oc which is negative (i.e. the mass is moving upwards) because Oc falls *on the tail portion* of the time-line.

Similarly, the line OD is drawn at π rad or 180° counterclockwise from OB and of length $\omega_n^2 X = A$, which is the maximum value of the acceleration. The instantaneous acceleration at time t, namely $a = A \cos(\omega_n t + \beta + \pi)$, is given by length Od which is negative (i.e. the mass is being retarded) because Od falls on the tail portion of the time-line.

Thus, the instantaneous values of displacement, velocity and acceleration are given by the lengths of Ob, Oc and Od which are, respectively, the values of the projections of their maximum-value vectors OB, OC and OD—each of which *remains fixed* on the phase diagram of Fig. 2.3(a) with phase differences of $\pi/2$ rad or 90° between them, and only the time-line moves at a constant rate of rotation (or angular frequency), ω_n. Such a polar or 'phase diagram', which gives a clear picture of the relative positions of the fixed X, V, and A vectors, enables the instantaneous value of x, v and a (in accordance with the position of the time-line at any time t) to be calculated merely from the geometry of a free-hand sketch without drawing it to scale and, of course, without further reference to the mathematical equations.

If the instantaneous values of the displacement are plotted as ordinates on a time base we get the familiar curve for undamped natural oscillations, i.e. for simple harmonic motion. The *non-dimensional* plot of the displacement equation, namely.

$\dfrac{x}{X} = \cos\left(2\pi\dfrac{t}{\tau_n} + \beta\right) = \cos 2\pi\left(\dfrac{t}{\tau_n} + \dfrac{\beta}{2\pi}\right)$, where τ is the periodic time $(2\pi/\omega_n)$, is shown in Fig. 2.3(b) in which $\beta = 0$ and $x = X$ when $t = 0$.

Similarly the non-dimensional versions of the velocity and acceleration equations are $\dfrac{v}{V} = \cos 2\pi\left(\dfrac{t}{\tau_n} + \dfrac{\beta}{2\pi} + \dfrac{1}{4}\right)$ and $\dfrac{a}{A} = \cos 2\pi\left(\dfrac{t}{\tau_n} + \dfrac{\beta}{2\pi} + \dfrac{1}{2}\right)$, respectively.

Phase difference can be represented by an interval along a time scale, but it is usually more helpful to compare this interval with the periodic time, e.g. as seen on Figs. 2.3(a) and (b) there is a lag of $\frac{1}{4}$ cycle between the maxima of displacement (X) and velocity (V), and also between velocity (V) and acceleration (A).

The same relative motion between vectors and time-line occurs, of course, if the latter is considered as fixed and all the vectors are imagined to rotate anti-clockwise in accordance with the alternative arbitrary convention usually used for alternating-current circuits in electrical engineering—the

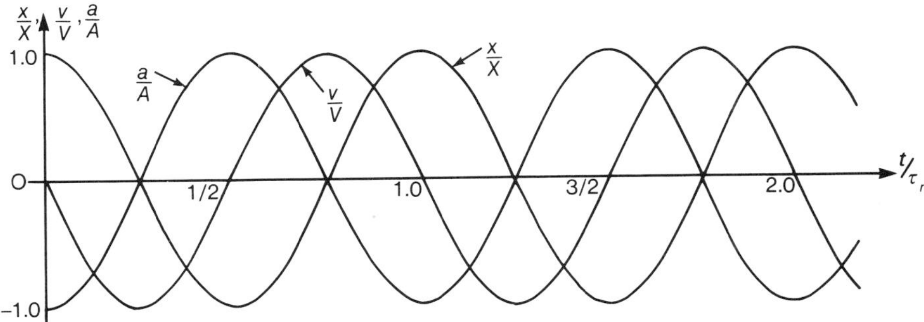

Fig. 2.3(b) Non-dimensional plot

horizontal (3-o-clock) line being taken as the reference for zero time. However, there is the advantage of easier interpretation of the phase diagram for mechanical vibrations if the several vectors representing elastic, damping, inertia and externally-applied forces remain fixed, and only the time-line is imagined as rotating as will be seen later in Figs. 3.3(d) and 4.4(b), etc.

EXAMPLE 2.3(i)—illustrative of the natural frequency of a mounted system
An instrument weighing 9.807 N is set on three rubber mounts, each rated of $\frac{1}{3}$ mm deflection per newton, i.e. the stiffness of each. Calculate the natural cyclic frequency of the system.

Stiffness of the three mounts together (i.e. of the system) is $k = 9$ N/mm. Mass of instrument is $m = W/g = \dfrac{9.807\text{ N}}{9.807\text{ m/s}^2}\left[\dfrac{\text{kg m}}{\text{N s}^2}\right] = 1$ kg. Hence, the natural frequency

$$\omega_n = \sqrt{(k/m)} = \sqrt{\left(\frac{9\text{ N}}{1\text{ kg mm}}\left[\frac{\text{kg }10^3\text{ mm}}{\text{N s}^2}\right]\right)} = \frac{95}{\text{s}}$$

or

$$\frac{95}{\text{s}}\left[\frac{\text{cycle}}{2\pi}\right] = 15.1\text{ cycle/s} = f_n$$

EXAMPLE 2.3(ii)—illustrative of oscillating masses on a taut wire
Determine the two natural frequencies and the ratio of the maximum vertical displacements or amplitudes of two equal masses each at distance L from the fixed ends of a horizontal taut wire of length $4L$. Assume that the increase in the tensile force, T, in the wire when in motion is negligible for small oscillations, and that the motions are simple harmonic in this system of two degrees of freedom.

Referring to Fig. 3.2(c) in which the positive direction of y is upwards, and assuming $y_1 = Y_1 \sin \omega t$ and $y_2 = Y_2 \sin \omega t$, the equations of motions of the two masses are, respectively:

$$m\ddot{y}_1 = -\frac{T}{L}y_1 - \frac{T}{2L}(y_1 - y_2) = -\frac{3T}{2L}y_1 + \frac{T}{2L}y_2$$

and

$$m\ddot{y}_2 = \frac{T}{2L}(y_1 - y_2) - \frac{T}{L}y_2 = \frac{T}{2L}y_1 - \frac{3T}{2L}y_2.$$

On substituting the harmonic equations, these become:

$$\left.\begin{aligned} -m_1Y_1\omega^2 + \frac{3T}{2L}Y_1 &= \frac{T}{2L}Y_2 \\ -m_2Y_2\omega^2 + \frac{3T}{2L}Y_2 &= \frac{T}{2L}Y_1 \end{aligned}\right\} \text{yielding } \frac{Y_1}{Y_2} = \frac{T/2L}{\left(\dfrac{3T}{2L} - m\omega^2\right)} \text{ and also } \frac{\left(\dfrac{3T}{2L} - m\omega^2\right)}{T/2L}$$

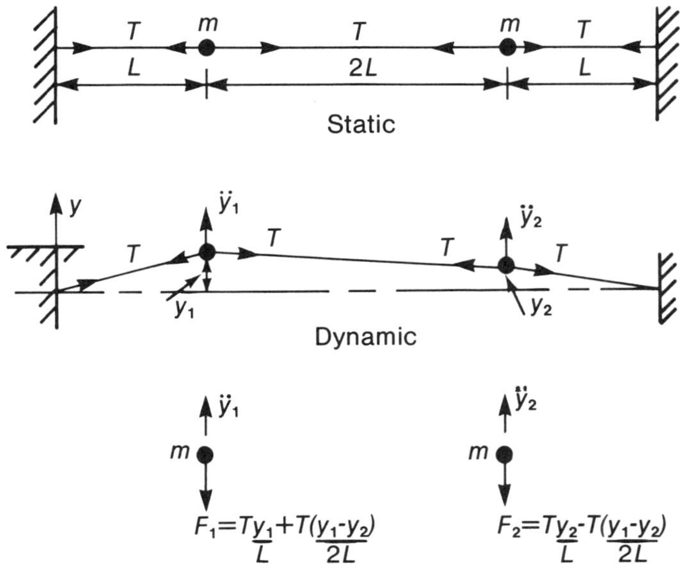

Fig. 2.3(c) Vibrating masses on a taut wire

from which we deduce that $\dfrac{3T}{2L} - m\omega^2 = \pm\dfrac{T}{2L}$ and, therefore, $m\omega^2 = \dfrac{T}{L}$ or $\dfrac{2T}{L}$. Hence, the two roots $\omega_1 = \sqrt{\dfrac{T}{mL}}$ and $\omega_2 = \sqrt{\dfrac{2T}{mL}}$ for which $Y_1 = +Y_2$ and $Y_1 = -Y_2$.

EXAMPLE 2.3(iii)—illustrative of natural frequency of torsional oscillations
A small electric generator is driven by the main shaft of an engine. The shaft is 10 mm diameter and 200 mm long having a modulus of rigidity 82.5 GN/m^2. If the moment of inertia of the rotor is 3.5 N mm s^2 (or 3.5 g m^2), estimate the natural cyclic frequency in torsion. State any assumptions made.

Assuming the moment of inertia of the engine is so large compared with that of the generator rotor, the engine end of the shaft may be assumed fixed relative to the rotor, and the system may be treated as a torsional pendulum composed of shaft and rotor. Hence, referring to Fig. 2.2(b), the natural cyclic frequency f_n is given by $\omega_n = 2\pi f_n$, where

$$\omega_n^2 = \frac{GJ}{LI} = \frac{\left(82.5 \times \dfrac{10^3\ \mathrm{N}}{\mathrm{mm}^2}\right) \times \left(\dfrac{\pi}{32} \times 10^4\ \mathrm{mm}^4\right)}{200\ \mathrm{mm} \times 3.5\ \mathrm{N\ mm\ s}^2} = \frac{11.56 \times 10^4}{\mathrm{s}^2}.$$

Thus, $\omega_n = 340/\text{s}$ – i.e. the natural frequency of torsional oscillation is $\omega_n = \frac{340}{\text{s}}\left[\frac{\text{Hz s}}{2\pi}\right] = 54.2$ Hz or $f_n = 54.2$ cycles/s.

EXAMPLE 2.3(iv)—illustrative of natural frequency of a mass on a lateral spring or beam

Deduce a formula for the undamped natural frequency of vertical oscillation of a mass *m* attached to the end of a cantilever beam.

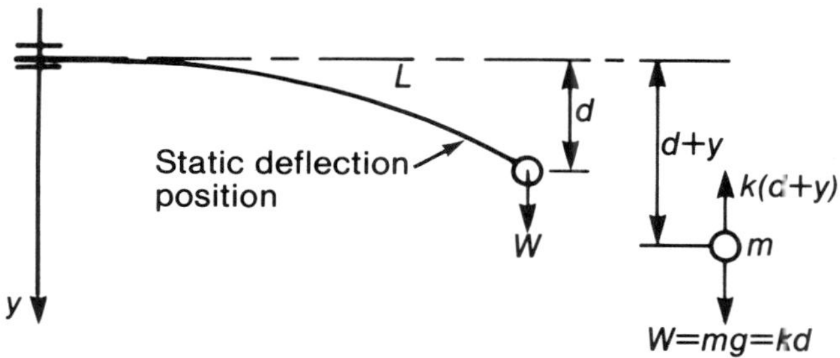

Fig. 2.3(d) Oscillating mass on a cantilever

Referring to Fig. 2.3(d), it will be appreciated that the 'spring force' may be exerted by any elastic structure—not, necessarily, a helical spring. In the static (or equilibrium) position, $d = \frac{WL^3}{3EI}$. Hence, the stiffness of the light (i.e. assumed of negligible mass) cantilever beam is $k = \frac{W}{d} = \frac{3EI}{L^3}$, and the differential equation of motion (as deduced from the free-body diagram) of the mass becomes $m\frac{d^2y}{dt^2} + \left(\frac{3EI}{L^3}\right)y = 0$ or $[D^2 + \omega_n^2]y = 0$, where $\omega_n^2 = \frac{3EI}{mL^3} = \frac{k}{m}$, and y is the downward displacement of mass m from the equilibrium position. As for Fig. 2.2(a), the solution of this differential equation gives $\tau_n = 2\pi/\omega_n$ and $f_n = \frac{\omega_n}{2\pi} = \frac{1}{2\pi}\sqrt{\left(\frac{3EI}{mL^3}\right)} = \frac{1}{2\pi}\sqrt{\left(\frac{g}{d}\right)}$.

EXAMPLE 2.3(v)—illustrative of oscillatory characteristics of a ship in rolling motion

Determine the periodic time.

Referring to Fig. 2.3(e), if I_o is the moment of inertia of the ship about the longitudinal rotational axis through O and the instantaneous angle of roll, θ, is small, then $T_o = I_o\ddot{\theta} = -Wh\theta$ (see Appendix A2) i.e. $\ddot{\theta} + Wh\theta/I_o = 0 = [D^2 + \omega_n^2]\theta$ and the periodic time of the rolling motion is $\tau_n = \frac{2\pi}{\omega_n} = 2\pi\sqrt{\left(\frac{I_o}{Wh}\right)} = 2\pi\sqrt{\left(\frac{k_o^2}{gh}\right)}$

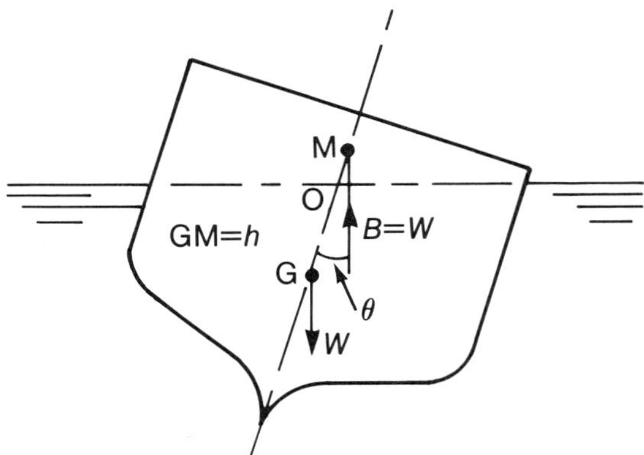

Fig. 2.3(e) Ship in rolling motion

The position of the rotational axis is usually unknown, and I_o is obtained from the period of roll of a model.

EXAMPLE 2.3(vi)—illustrative of torsional oscillations and a node.
Determine the period of the undamped natural torsional vibrations of a shaft on which there are two rotating masses.

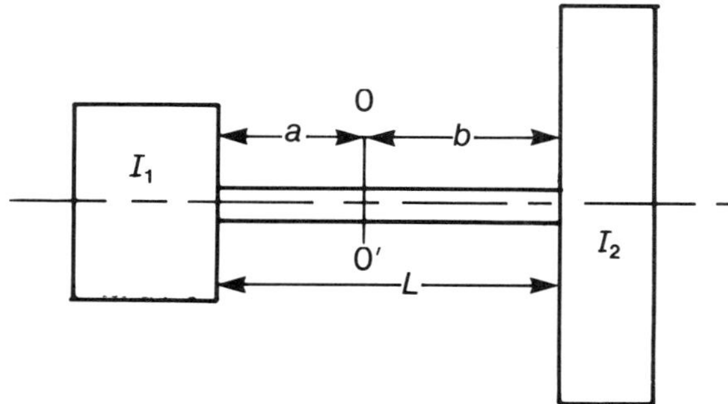

Fig. 2.3(f) Torsional oscillation of a shaft

If the system is as shown diagrammatically in Fig. 2.3(f) and the effect of the mass of the shaft is neglected, the discs will oscillate in opposite directions so that the angular momentum of the system with respect to the axis of the shaft will always be zero, i.e. the clockwise angular momentum of one will equal the anticlockwise momentum of the other at any instant in time. Also, during such oscillations a certain cross-section (the nodal section) oo′ will remain unmoved, and its position is found from the condition that the parts to the right and left of it have the same periodic time, namely, $\tau_n = 2\pi\sqrt{\left(\frac{aI_1}{GJ}\right)} = 2\pi\sqrt{\left(\frac{bI_2}{GJ}\right)}$. Hence, $aI_1 = bI_2$ or $\frac{a}{b} = \frac{I_2}{I_1}$ and $\frac{a+b}{a} = \frac{I_1+I_2}{I_1} = \frac{L}{b}$. Therefore, the periodic time of the undamped natural (or free) torsional vibration is $\tau_n = 2\pi\sqrt{\left(\frac{I_1 I_2 L}{GJ(I_1+I_2)}\right)}$ as, also, found in Section 2.6.

EXAMPLE 2.3(vii)—illustrative of natural oscillations in torsion, and locating a node
A steel shaft 50 mm diameter and 0.6 m long is attached at one end to a flywheel of mass 180 kg and radius gyration 0.25 m. The other end of the shaft is attached to an armature, the moment of inertia of which is 0.6 that of the flywheel. Calculate the natural cyclic frequency of torsional oscillation, and locate the node. $G = 82.5\ \text{GN/m}^2$.

Moment of inertia of flywheel is $I_2 = m_2 k_2 = 180\ \text{kg} \times 0.0625\ \text{m}^2 = 11.25\ \text{kg m}^2$. The polar second moment of area of the shaft is $J = \frac{\pi}{32} d^4 = \frac{\pi}{32} \frac{625}{10^8}\ \text{m}^4$. Hence,

$$f_n = \frac{1}{\tau_n} = \frac{1}{2\pi} \sqrt{\left(\frac{GJ(1 + I_2/I_1)}{I_2 L} \right)}$$

$$= \frac{1}{2\pi} \sqrt{\left(82.5 \times 10^9 \frac{\text{N}}{\text{m}^2} \times \frac{\pi}{32} \frac{625\ \text{m}^4}{10^8} \frac{(1 + 1/0.6)}{11.25\ \text{kg m}^2 \times 0.6\ \text{m}} \right)}$$

i.e.

$$f_n = \frac{1}{2\pi} \sqrt{\left(2 \times 10^4 \frac{\text{N}}{\text{kg m}} \right)} = \frac{141.4}{2\pi\ \text{s}} = 22.5\ \text{cycle/s or } 22.5\ \text{Hz}.$$

and

$$b = \frac{L}{1 + I_2/I_1} = \frac{0.6\ \text{m}}{1 + 1/0.6} = \frac{0.36}{1.6}\ \text{m} = 225\ \text{mm}.$$

EXAMPLE 2.3(viii)—illustrative of vibrations in a spring loaded governor
A centrifugal governor is of the type in which a pair of masses, connected by a helical spring, slide equally in opposite radial directions. At 1200 rev/min the radius of the masses is 100 mm, and at 1320 rev/min, 115 mm. Calculate the frequency of natural oscillation of the governor when rotating at 1200 rev/min.

Referring to Fig. 2.3(g), the increase in centrifugal force on the spring when the masses are displaced by δr is $m\omega^2 \delta r$, and the increase in restoring force is $k\,\delta r$, where k is the stiffness of half the spring (see Appendix A8). Hence, the net outward force is $-k\delta r + m\omega^2 \delta r = ma$, and the acceleration along the radius is $a = -(k/m - \omega^2)\delta r$, i.e. the acceleration, a, is negatively proportional to displacement δr. Hence, the oscilla-

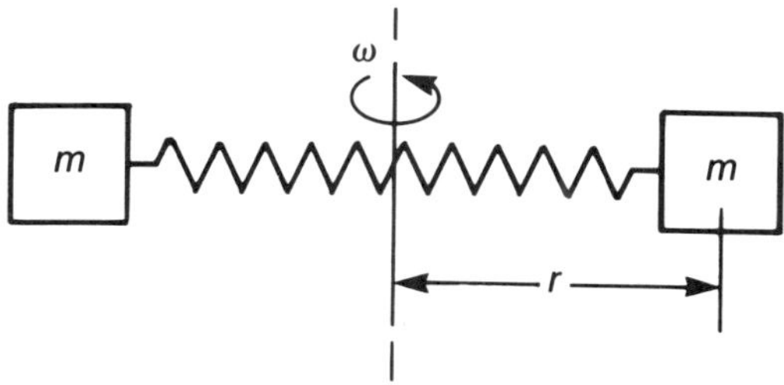

Fig. 2.3(g) Centrifugal governor

tory motion along the radius is simple harmonic, and of angular frequency $\omega_n = \sqrt{(k/m - \omega^2)} = \sqrt{(-a/\delta r)}$.

In particular, since $k \times 15\text{ mm} = m(r_2\omega_2^2 - r_1\omega_1^2)$,

$$\frac{k}{m} = \frac{1}{15\text{ mm}}\left\{115\text{ mm}\left(22\frac{\text{rev}}{\text{s}}\right)^2 - 100\text{ mm}\left(20\frac{\text{rev}}{\text{s}}\right)^2\right\}\left[\frac{4\pi^2}{\text{rev}^2}\right]$$

$$= \frac{4\pi^2}{3}\{23 \times 484 - 20 \times 400\}\ 1/\text{s}^2 = 41\ 250/\text{s}^2$$

Hence, the angular frequency of natural oscillations at 1200 rev/min is $\omega_n = \sqrt{\left\{\frac{41250 - (20 \times 2\pi)^2}{\text{s}^2}\right\}} = 159/\text{s}$ and the cyclic frequency, f_n, is $\frac{159}{\text{s}}\left[\frac{\text{cycle}}{2\pi}\right] =$ 25.3 cycle/s or 25.3 Hz.

EXAMPLE 2.3(ix)—illustrative of natural frequency of oscillation

A loaded trailer of mass 4500 kg is hitched by means of a strong helical spring of stiffness 175 N/mm to a car of mass 1500 kg. Calculate the frequency of free oscillations as the car moves off, assuming harmonic motion in the spring.

Referring to Fig. 2.3(h), the extension of the spring is $(x_1 - x_2)$ caused by a force $F = k(x_1 - x_1)$.

Hence, the laws of motion of the two masses are:

$$m_1\ddot{x}_1 = -k(x_1 - x_2)$$

and

$$m_2\ddot{x}_2 = +k(x_1 - x_2)$$

which, on substituting the harmonics $x_1 = X_1 \sin\omega_n t$ and $x_2 = X_2 \sin\omega_n t$, become the double algebraic equation:

$$m_1\omega_n^2 X_1 = k(X_1 - X_2) = -m_2\omega_n^2 X_2.$$

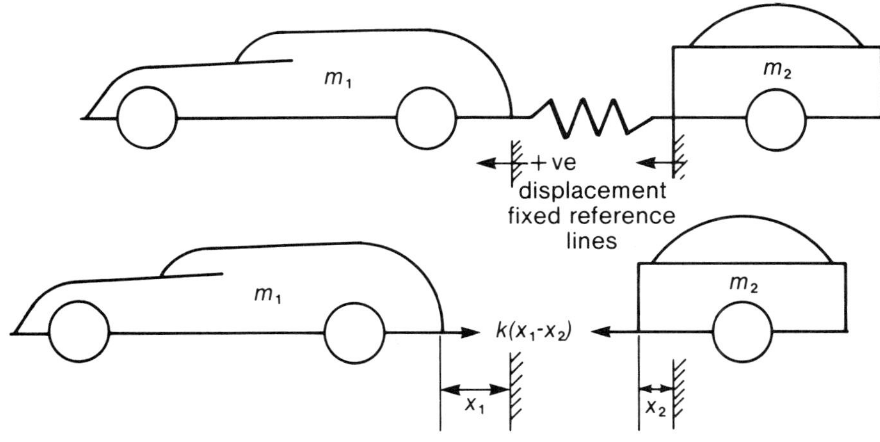

Fig. 2.3(h) Loaded trailer

Hence,

$$\frac{X_1}{X_2} = -\frac{m_2}{m_1} = -3$$

and

$$\omega_n^3 = \frac{k}{m_1}\left(1 - \frac{X_2}{X_1}\right) = \frac{175\,\text{N}}{1500\,\text{kg mm}}\left(1 + \frac{1}{3}\right)\left[\frac{\text{kg } 10^3\,\text{mm}}{\text{N s}^2}\right]$$

$$= \frac{175}{1.5} \times \frac{4}{3} \times \frac{1}{\text{s}^2} = 155.5/\text{s}^2$$

and the frequency of oscillations in the line of motion is $\omega_n = \frac{12.47}{\text{s}}\left[\frac{\text{cycle}}{2\pi}\right] = 1.98\,\text{cycle/s}$

There would, of course, be zero frequency if $(x_1 - x_2) = 0$, i.e. if the system were rigid.

2.4 NATURAL FREQUENCY ESTIMATION USING ENERGY METHODS

Oscillatory and vibration problems which do not conveniently lend themselves to direct application of Newton's 2nd law of motion may be solved by the energy method already introduced in Section 2.2; and it is sometimes advantageous to set up the differential equations of motion via energy considerations—i.e. by applying the law of conservation of energy (see Section 1.4). For example, in the case of oscillating liquids, the restoring force is caused by gravity, e.g. the motion of the liquid in the uniform U tube of Fig. 2.4(a) is neither simple translational nor rotational about a fixed axis.

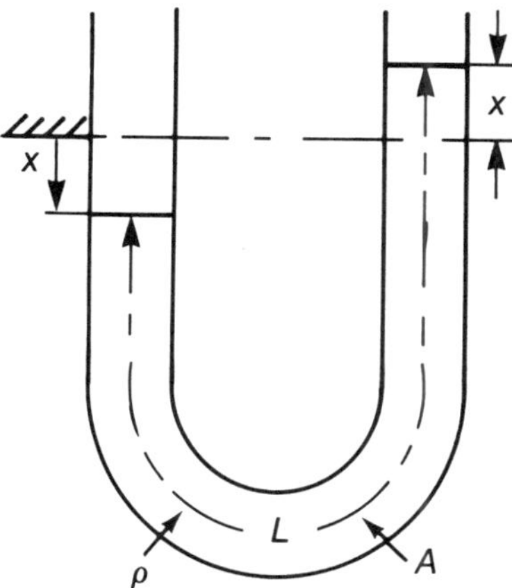

Fig. 2.4(a) Oscillating liquid

It is, therefore, advantageous to use an energy method for finding frequency from the fact that the kinetic energy of the liquid at any instant is $T = \frac{1}{2}(\rho AL)\dot{x}^2$ and the corresponding potential energy is $V = \rho gAx^2$. The latter because the work done on the liquid is the same as if the column of length x had been transferred from the left leg to the right leg, namely, $-\rho gAx^2$ (i.e. to lift this liquid a distance x—the positive direction being downwards). Hence, since $\frac{\mathrm{d}}{\mathrm{d}t}(T + V) = 0$ for a conservative system, we deduce that

$$\rho AL\dot{x}\ddot{x} + 2\rho gAx\dot{x} = 0$$

i.e.

$$\ddot{x} + 2gx = 0 = [\mathrm{D}^2 + \omega_{\mathrm{n}}^2]x, \text{ where } \omega_{\mathrm{n}} = \sqrt{(2g/L)}$$

and the periodic time is $\tau_{\mathrm{n}} = 2\pi\sqrt{\left(\frac{L}{2g}\right)}$ which is independent of A, ρ, and the shape of the uniform tube.

Similarly, the effect of the mass of a spring can be taken into account relatively easily by using the fact that $T + V$ remains constant for a conservative (i.e. undamped) system. Thus, considering the 'massive' spring of Fig. 2.4(b), the kinetic energy of the element δz is $\frac{1}{2}\left(\frac{w}{g}\delta z\right)\left(\frac{z}{L}\dot{x}\right)^2$, where w is the weight of spring per unit length. Hence, the total kinetic energy of the spring at any instant is

$$\frac{1}{2}\frac{w\dot{x}^2}{gL^2}\int_0^L z^6\,\mathrm{d}z = \frac{1}{3}\left\{\frac{1}{2}\frac{w}{g}L\dot{x}^2\right\}.$$

Considering now the mass M of weight W, its kinetic energy at that same instant is $\frac{1}{2}\frac{W}{g}\dot{x}^2$. Hence, the total *kinetic* energy of the spring-mass system is $T = \frac{1}{2}\left(\frac{W}{g} + \frac{1}{3}\frac{wL}{g}\right)\dot{x}^2$, from which it will be seen that, to include the effect of the mass of the spring on the periodic time of natural vibration of the system, $\frac{1}{3}$ the mass of the spring must be added to the mass of the oscillating body M. The potential energy $V = \frac{1}{2}kx^2$, as proved in Section 2.2. Thus, the total energy of the spring-mass system of Fig. 2.4(b) at any instant is $T + V = \left\{\frac{1}{2}\left(\frac{W}{g} + \frac{1}{3}\frac{wL}{g}\right)\dot{x}^2\right\} + \{\frac{1}{2}kx^2\} =$ constant. Comparing this equation with that corresponding in Section 2.2, we may say that the period of vibration

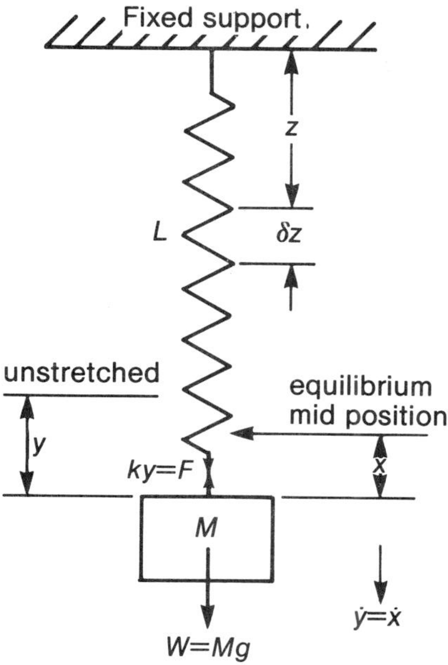

Fig. 2.4(b) Effect of massive spring

will be the same as for a spring assumed massless (Fig. 2.2(a)) but supposing a mass $\left(M + \dfrac{1}{3}\dfrac{wL}{g}\right) = (M + m/3)$. Hence, $\omega_n = \sqrt{\left(\dfrac{k}{M + m/3}\right)}$ and $\tau_n = 2\pi\sqrt{\left(\dfrac{M + m/3}{k}\right)}$.

This result can also be deduced from the other aspect of the energy method, namely, $T_{max} = V_{max}$. For, if X is the maximum amplitude then $V_{max} = \frac{1}{2}kX^2$ and, assuming harmonic motion $x = X\cos\omega_n t$ the kinetic energy of the spring-mass system in mid-position is $T_{max} = \frac{1}{2}M\omega_n^2 X^2 + \dfrac{1}{2}\dfrac{m}{3}\omega_n^2 X^2$. Hence, $k = \left(M + \dfrac{m}{3}\right)\omega_n^2$ as before.

EXAMPLE 2.4(i)—illustrative of an energy method for finding natural frequency
Equating the maxima of kinetic and potential energies, find the period of oscillation of fluid in the tank system shown diagrammatically in Fig. 2.4(c). State any assumptions made.

Assumptions:
(i) the fluid is incompressible,
(ii) damping is negligible—i.e. the fluid is inviscid,
(iii) the motion is simple harmonic (verifiable in practice).

Referring to Fig. 2.4(c), the volume $A_1 Y_1 = A_2 Y_2 = aY$, where the Y's are maxima.

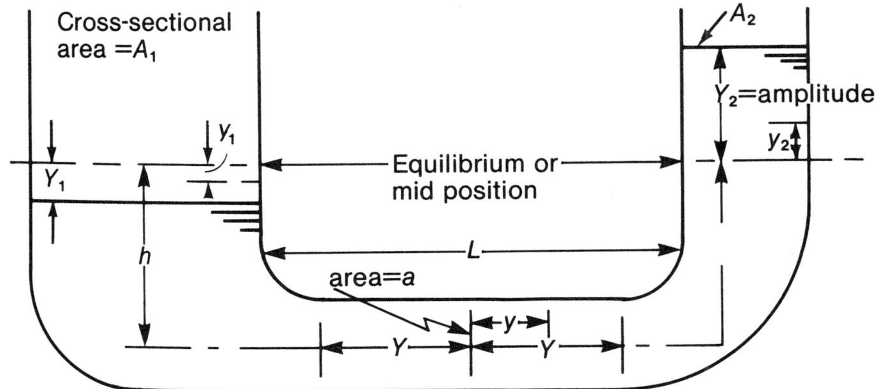

Fig. 2.4(c) Undamped natural oscillation of liquid

Also, $A_1 y_1 = A_2 y_2 = ay$, where the y's are corresponding displacements from respective equilibrium positions.

Since $V_{max} = T_{max}$, i.e. the potential energy in the extreme position is equal to the kinetic energy in the mid-position, then

$$\rho g A_1 Y_1 \left(\frac{Y_1 + Y_2}{2} \right) = \tfrac{1}{2}\rho A_1 h(\dot{y}_1)^2_{max} + \tfrac{1}{2}\rho A_2 h(\dot{y}_2)^2_{max} + \tfrac{1}{2}\rho a L(\dot{y})^2_{max}$$

Hence,

$$ga^2 Y^2 \left(\frac{A_1 + A_2}{A_1 A_2} \right) = h\frac{a^2}{A_1}\dot{y}^2_m + h\frac{a^2}{A_2^2}\dot{y}^2_m + h\frac{a^2}{a}\dot{y}^2_m$$

or

$$gY^2 \left(\frac{A_1 + A_2}{A_1 A_2} \right) = \dot{y}^2_m \left(\frac{h}{A_1} + \frac{h}{A_2} + \frac{L}{a} \right).$$

Thus, since the motion will be simple harmonic, i.e. $y = Y \cos \omega_n t$, then $\dot{y} = -\omega_n Y \sin \omega_n t = -\omega_n Y$ in mid-position ($\omega_n t = \pi/2$) and

$$gY^2 \left(\frac{A_1 + A_2}{A_1 A_2} \right) = \omega_n^2 Y^2 \left(\frac{h}{A_1} + \frac{h}{A_2} + \frac{L}{a} \right)$$

which yields

$$\omega_n = \sqrt{\left\{ \frac{g\left(\dfrac{1}{A_1} + \dfrac{1}{A_2} \right)}{h\left(\dfrac{1}{A_1} + \dfrac{1}{A_2} \right) + \dfrac{L}{a}} \right\}} \text{ and } \tau = 2\pi/\omega_n.$$

Compare these results with those of the uniform tube of Fig. 2.4(a).

EXAMPLE 2.4(ii)—illustrative of the energy method applied to lateral vibrations
Find the effective mass influencing the period of vibration of a simply-supported beam whose mass per unit length is m' carrying a heavy central load $W = Mg$.

Referring to Fig. 2.4(d) and assuming the mass $m'L$ of the beam is small compared with the central mass M, the deflection curve of the beam during vibration can be assumed to have the same shape as the statical deflection curve caused by the concentrated load W with reactions $W/2$. The static deflection curve can be deduced by integrating the bending moment equation (see Appendix A4):
$EI\dfrac{d^2y}{dx^2} = -\dfrac{W}{2}x$ at any section x. This results in $EIy = \dfrac{W}{48}(3L^2x - 4x^3)$, and $Y = \dfrac{WL^3}{48EI}$ is the static deflection at mid-span. Hence, $y = \left(\dfrac{3x}{L} - 4\dfrac{x^3}{L^3}\right)Y$. The velocity of the element of mass $m\delta x$ is $\dot{y} = \left(\dfrac{3x}{L} - \dfrac{4x^3}{L^3}\right)\dot{Y}$ from which, as expected, when $x = L/2$ then $\dot{y} = \dot{Y}$. Therefore, the kinetic energy of the beam at any instant is

$$T_B = 2\int_0^{L/2}\tfrac{1}{2}(m'\delta x)\left\{\left(\frac{3x}{L} - \frac{4x^3}{L^3}\right)\dot{Y}\right\}^2 = m'\dot{Y}^2 - \int_0^{L/2}\left(\frac{3x}{L} - \frac{4x^3}{L^3}\right)d\left(\frac{x}{L}\right) = \frac{17}{35} - \frac{m'}{2}L\dot{Y}^2,$$

and the kinetic energy of M is $T_M = \frac{1}{2}M\dot{Y}^2$.

Hence, the total kinetic energy of the system at any instant is $T = T_B + T_M = \frac{1}{2}\left(M + \dfrac{17}{35}m'L\right)\dot{Y}^2$, i.e. the effective mass is $\left(M + \dfrac{17}{35}m'L\right)$ and the period of vibration will be the same as for a massless beam loaded at mid-span but with a mass $\left(M + \dfrac{17}{35}m'L\right)$ instead of M, namely, $\tau_n = 2\pi\sqrt{\left\{\left(M + \dfrac{17}{35}m'L\right)\left(\dfrac{L^3}{48EI}\right)\right\}}$ instead of $2\pi\sqrt{\left(\dfrac{ML^3}{48EI}\right)}$—(also see Exercise No. 5 at end of Chapter 2).

Alternatively, since Rayleigh's method does not require that we, necessarily, assume the static deflection curve but any reasonable shape that satisfies the boundary conditions, we may assume $y = Y\sin \pi x/L$ which results in an effective mass $\left(M + \dfrac{17}{34}m'L\right)$—see also, Example 2.5(i).

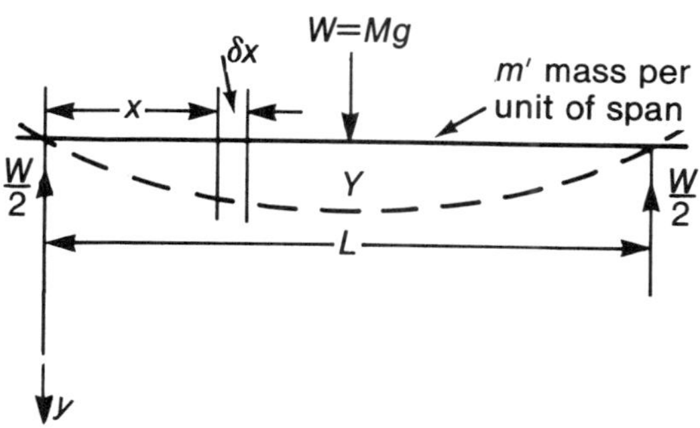

Fig. 2.4(d) Lateral vibration of a loaded beam

EXAMPLE 2.4(iii)—illustrative of quantitative estimation of lateral natural frequency
If in Example 2.4(ii) $M = 6.5$ kg, and the beam is steel of dimensions 1.5 m × 72 mm × 10 mm, calculate the natural frequency of vibration taking $E = 200\ \text{GN/m}^2$ and steel of density 8 Mg/m^3.

The second moment of area of the beam of rectangular cross section is

$$I = \frac{1}{12}bd^3 = \frac{1}{12} \times 72\ \text{mm} \times 10^3\ \text{mm}^3 = 6000\ \text{mm}^4.$$

$$m'L = \frac{1.5\ \text{m} \times 720\ \text{m}^2}{10^6} \times \frac{8\ \text{Mg}}{\text{m}^3} = 8.64\ \text{kg}$$

Hence, the frequency

$$\omega_n = \sqrt{\left\{\left(\frac{48EI}{M + 0.486m'L}\right)\frac{1}{L^3}\right\}}$$

$$= \sqrt{\left\{\frac{48 \times 2 \times 10^{11}\dfrac{\text{N}}{\text{m}^2} \times \dfrac{6}{10^9}\text{m}^4\left[\dfrac{\text{kg m}}{\text{N s}^2}\right]}{(6.5 + 4.2)\text{kg} \times (1.5)^3\ \text{m}^3}\right\}}$$

$$= \sqrt{\left\{\frac{576 \times 10^2}{36.2} \times \frac{1}{\text{s}^2}\right\}} = \frac{39.9}{\text{s}}\left[\frac{\text{cycle}}{2\pi}\right] = 6.36\ \text{cycle/s}$$

2.5 RAYLEIGH'S ENERGY METHOD OF ESTIMATING NATURAL FREQUENCY

In applying the energy method ($T_{max} = V_{max}$) to systems with distributed or continuous masses—e.g. vibrating beams (which are multi or continuous-mass lateral or transversely vibrating systems) the deflection configuration needs to be known or assumed in order that the kinetic and potential energies may be determined. Rayleigh showed that the fundamental frequency of such systems can be found with good accuracy by assuming the static deflection curve (or any reasonable shape that satisfies the boundary conditions) of the elastic beam to be the true curve or mode-shape of the system,—also assumed vibrating with s.h.m.

Although Rayleigh's energy method is applied to systems of more than one degree of freedom and is relatively easy to apply—being less laborious than the more accurate second law of motion or momentum application, only the fundamental frequency can be found thereby. Also, such a calculated frequency can never be lower than the true fundamental natural frequency of the system. The potential (i.e. strain) energy in a deflected beam is a function of its deflected shape—unaffected by the gravitational field. Thus, the general method for determining the fundamental frequencies of beams involves the

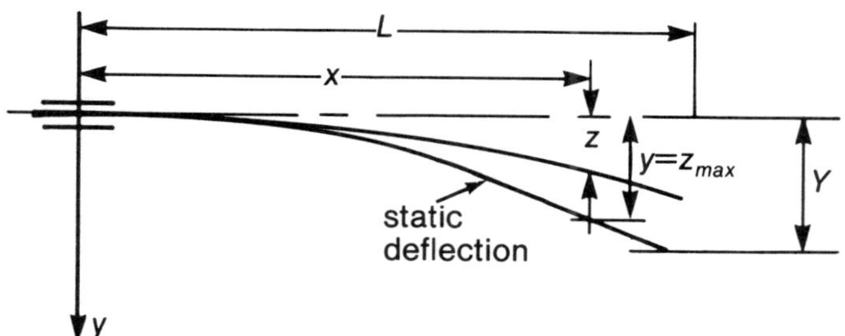

Fig. 2.5(a) Vibrating beam

assumptions of (i) a mode shape, (ii) harmonic motion, (iii) a conservative system, and may be developed as follows. Let m be the total ($m = m'L$) mass distribution along the beam of cross-sectional area a—i.e. $\delta m = \rho a \delta x$ (corresponding to a loading intensity or weight per unit length $= w = g\rho a = m'g$). Let y be the deflection (i.e. amplitude) at any point x along the elastic (static deflection) curve corresponding to the fundamental mode of vibration shown diagrammatically in Fig. 2.5(a) where z is the deflection at any time t of point x vibrating is simple harmonic motion according to the law $z = y \cos \omega_n t$.
Hence, we may deduce that $\dot{z} = -\omega_n y \sin \omega_n t$

$$\dot{z}_{max} = -\omega_n y \text{ when } \omega_n t = \frac{\pi}{2} \text{ or } t = \frac{\tau_n}{4},$$

and the maximum kinetic energy of the beam is

$$T_{max} = \sum_0^L \tfrac{1}{2}\delta m(\dot{z}_{max})^2 = \tfrac{1}{2}\omega_n^2 \int_0^L y^2\, dm, \quad \text{where } \delta m = m'\delta x$$

The 'potential' energy of the beam is determined by the work done on the beam and which is stored as strain energy, namely,

$$V_{max} = \tfrac{1}{2} EI \int_0^L \left(\frac{d^2 y}{dx^2}\right)^2 dx$$

Equating the two maxima yields

$$\omega_n^2 = \frac{EI}{m'} \frac{\int_0^L \left(\frac{d^2 y}{dx^2}\right)^2 dx}{\int_0^L y^2\, dx} \quad \text{and} \quad \tau_n = 2\pi/\omega_n.$$

This calculated ω_n will be a few per cent high because the actual deflection curve is caused by inertia forces rather than by static loading (see also, Section 6.2).

If a beam can be assumed to carry several concentrated masses ($M_1, M_2, \ldots$) then, as a first approximation, the maximum potential (strain) energy can be taken as the work done by the weights of the masses during static deflection ($y_1, y_2, \ldots$) of the beam, namely, $V_{max} = \sum\left(\frac{M}{2}gy\right)$ and the maximum kinetic energy $T_{max} = \sum(\frac{1}{2}\omega^2 My^2)$ from which we deduce that $\omega^2 = \frac{g\sum(My)}{\sum(My^2)}$.

EXAMPLE 2.5(i)—illustrative of frequency estimation by equating maximum kinetic and potential energies

Estimate the fundamental natural frequency of a uniformly-loaded cantilever beam—the mass per unit length being $m' = \rho a$, where a is the cross-sectional area, and ρ the density of the material.

Referring to Fig. 2.5(a), we have a continuous elastic body with a large number of degrees of freedom but, by assuming the form of deflection curve or shape, Rayleigh's method enables the system to be regarded as having one degree of freedom. Thus, assuming the deflection-shape of the beam to be given by $y = \frac{1}{2}Y\left\{3\left(\frac{x}{L}\right)^2 - \left(\frac{x}{L}\right)^3\right\}$ which satisfies the boundary conditions $x = 0, y = 0; x = 0, \frac{dy}{dx} = Y\left\{6\left(\frac{x}{L^2}\right) - 3\left(\frac{x^2}{L^3}\right)\right\} = 0$, i.e. zero slope and, since bending moment $M \propto \frac{d^2y}{dx^2}$ (see Appendix A4) and since $\frac{d^2y}{dx^2} = Y\left\{\frac{6}{L^2} - \frac{6x}{L^3}\right\}$ then when $x = L, M = 0$ which satisfies the boundary conditions $x = L$, $M = 0$, this shape satisfies the two geometric boundary conditions. Thus, we may proceed (since $\delta m = m'dx$) to deduce that

$$T_{max} = \frac{\omega_n^2}{2}\int_0^L y^2 \, dm = \frac{m'\omega_n^2}{8}YL^2\int_0^L \left\{3\left(\frac{x}{L}\right)^2 - \left(\frac{x}{L}\right)^3\right\}^2 d\left(\frac{x}{L}\right)$$

$$= \frac{33}{280}\rho a\omega_n^2 YL$$

and

$$V_{max} = \frac{EI}{2}\int_0^L \left(\frac{d^2y}{dx^2}\right)^2 dx = \frac{EI}{8}Y^2\int_0^L \left(\frac{6}{L^2} - \frac{6x}{L^3}\right)^2 dx = \frac{3EI}{2L^3}Y^2.$$

Hence, equating T_{max} to V_{max} yields $\omega_n^2 = \frac{12.73}{m'}\frac{EI}{L^4} = 12.73\frac{EI}{\rho aL^4}$ and frequency $f = \frac{1}{\tau_n} = \frac{\omega_n}{2\pi} = 0.566\sqrt{\left(\frac{EI}{\rho aL^4}\right)}$.

Alternatively, we could have assumed the shape $y = Y(1 - \cos(\pi x/2L)$ fcr which $\dfrac{dy}{dx} = Y\left(\dfrac{\pi}{2L}\sin\left(\dfrac{\pi x}{2L}\right)\right)$ which satisfy the conditions $x=0$, $y=0$ and $\dfrac{dy}{dx}=0$ (zero slope); and since $\dfrac{d^2y}{dx^2} = Y\left(\dfrac{\pi}{2L}\right)^2 \cos\left(\dfrac{\pi x}{2L}\right)$, the other boundary condition $x=0$ and $M=0$ is satisied. This shows that the latter shape could be used equally well as the former.

EXAMPLE 2.5(ii)—illustrative of Rayleigh's method
Determine the fundamental frequency of lateral vibration of a simply-supported beam of cross-sectional area a and density ρ, i.e. $\delta m = m'\delta x = \rho a \delta x$.

The deflection curve is represented diagrammatically on Fig. 2.5(b) and $y = Y\sin(\pi x/L)$ is the correct shape formula without assumptions and from which we deduce that $\dfrac{dy}{dx} = Y\dfrac{\pi}{L}\cos\left(\dfrac{nx}{L}\right)$ and $\dfrac{d^2y}{dx^2} = -Y\dfrac{\pi^2}{L^2}\sin\left(\dfrac{\pi x}{L}\right)$. Hence, the geometric boundary conditions $x=0$, $y=0$ and $x=L$, $y=0$; $x=0$, $M=0$, and $x=L$, $M=0$; also, $x=\dfrac{L}{2}, \dfrac{dy}{dx}=0$ (i.e. zero slope at mid-span) are satisfied. Thus, as expected, the shape satisfies the conditions, and since $\delta m = \rho a \delta x$

$$T_{max} = \tfrac{1}{2}\omega_n^2 \int_0^L y^2\, dm = \frac{\rho a}{2}\omega_n^2 \int_0^L y^2\, dx$$

$$= \frac{\rho a}{2}\omega_n^2 Y^2 \int_0^L \sin^2\left(\frac{\pi x}{L}\right) dx$$

and

$$V_{max} = \frac{EI}{2}\int_0^L \left(\frac{d^2y}{dx^2}\right)^2 dx = \frac{\pi^4}{L^4} Y^2 \frac{EI}{2}\int_0^L \sin^2\left(\frac{\pi x}{L}\right) dx$$

Equating T_{max} to V_{max} yields $\omega_n^2 = \dfrac{\pi^4}{L^4}\dfrac{EI}{\rho a}$ and $f = \dfrac{\omega_n}{2\pi} = \dfrac{\pi}{2L^2}\sqrt{\left(\dfrac{EI}{\rho a}\right)}$

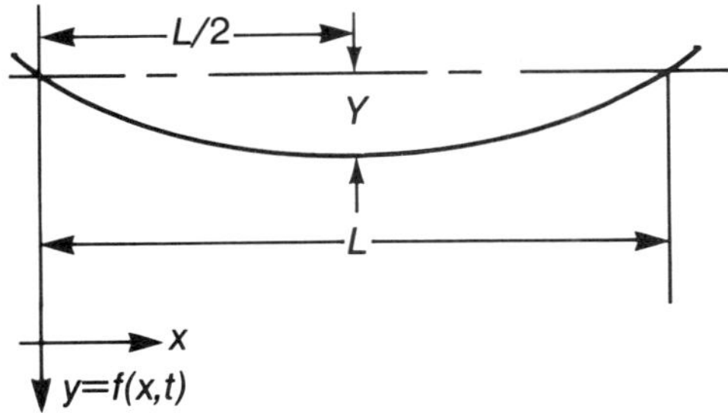

Fig. 2.5(b) Lateral vibration

EXAMPLE 2.5(iii)—illustrative of the estimation of natural frequency by an energy method

Assuming harmonic motion, determine the fundamental frequency of a uniformly-loaded cantilever beam with a concentrated mass M at the end, equal to the mass of the uniformly distributed load.

$$V_{\max} = \frac{3}{2}\frac{EI}{L^3} Y^2 \text{ from Example 2.5(i), and } T_{\max} = \frac{33}{280} M\omega_n^2 Y^2 + \tfrac{1}{2} M\omega_n^2 Y^2.$$

Equating these maxima yields

$$\frac{3}{2}\frac{EI}{L^3} = M\omega_n^2\left(\frac{33}{280} + \frac{1}{2}\right)$$

and

$$\omega_n^2 = \frac{3 \times 140}{173}\frac{EI}{ML^3} = 2.43\frac{EI}{ML^3}.$$

(see, also, Dunkerley's method in Example 6.5(ii)).

2.6 UNDAMPED NATURAL OSCILLATIONS WITH SEVERAL DEGRESS OF FREEDOM

Whilst it is possible in many cases to simplify the dynamics of complex systems and use one degree-of-freedom as a first approximation, there are cases when this cannot be done—e.g. when there is more than one mass or when there is rotational as well as translational motion in a system. On the other hand, the amplitude of vibrations of a system with one degree-of-freedom can be reduced in a vertical plane by attaching a dynamic neutralizer (see Fig. 5.1(a)) which results in a system with two degrees-of-freedom.

A common practical problem with two degrees of freedom is that of torsional vibrations—e.g. in motor-generator sets, motor-driven pumps and fans, blade discs in jet engines. The representative case for analysis is shown in Fig. 2.6(a)—the discs of moment of inertia I_1 and I_2 being keyed to the ends of a uniform shaft of torsional stiffness $k = \dfrac{GJ}{L}$. Thus, assuming negligible damping, the angle of twist of the shaft $\theta_1 - \theta_2$, causes a torque on disc 1 of $-k(\theta_1 - \theta_2) = I_1\ddot{\theta}_1$ and on disc 2 of $+k(\theta_1 - \theta_2) = I_2\ddot{\theta}_2$. Hence, assuming harmonic motion in a principal mode of angular frequency ω, namely, $\theta_1 = \Theta_1 \cos\omega t$ and $\theta_2 = \Theta_2 \cos\omega t$ the two equations of motion yield the two simultaneous algebraic equations

$$k\Theta_2 = \Theta_1(k - \omega^2 I_1)$$

and

$$k\Theta_1 = \Theta_2(k - \omega^2 I_2)$$

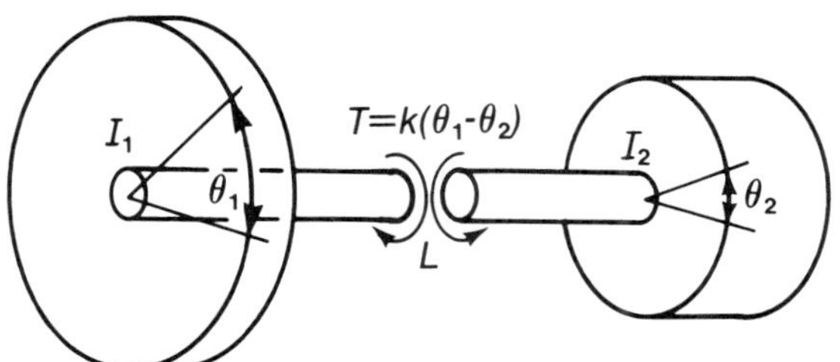

Fig. 2.6(a) Torsional oscillation

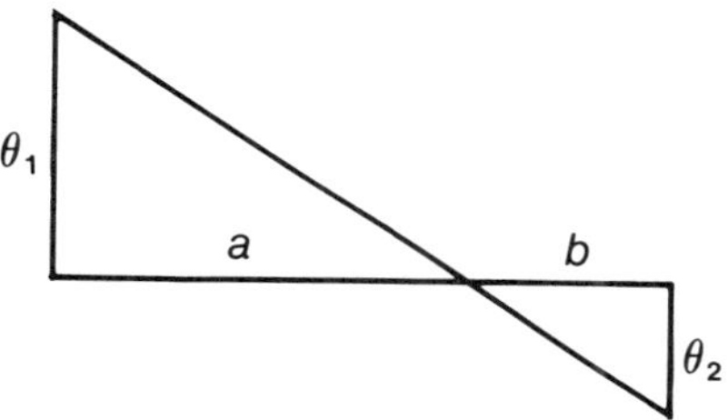

Fig. 2.6(b) Position of a node

from which we deduce that $\dfrac{\Theta_L}{\Theta_2}=\dfrac{k}{k-\omega^2 I_1}=\dfrac{k-\omega^2 I_2}{k}$ and the frequency equation $\omega^2\{I_1I_2\omega^2-k(I_1+I_2)\}=0$. Hence, the two solutions are $\omega=0$ (indicating zero twisting of the shaft) and $\omega=\sqrt{\left\{\dfrac{k(I_1+I_2)}{I_1I_2}\right\}}$ with corresponding maximum amplitude ratios of $\dfrac{\Theta_1}{\Theta_2}=1$ and $-\dfrac{I_2}{I_1}$, respectively. Thus, at any time t, we deduce that either $\theta_1=\theta_2$ indicating no twist, or $\theta_1 I_1=-\theta_2 I_2$ indicating that, when the shaft twists, in this particular principal mode the discs oscillate in opposite directions with a node in the shaft given by $\dfrac{\Theta_1}{-\Theta_2}=\dfrac{a}{b}=\dfrac{I_2}{I_1}$ as shown diagrammatically in Fig. 2.6(b) and previously used in the solution of Example 2.3(vi). The solution $\omega=0$ and ratio $\theta_1/\theta_2=1$ means that the shaft does not twist but executes rigid-body motion with an infinitely long period, i.e. with no friction, rotation would continue endlessly!

EXAMPLE 2.6(i)—illustrative of torsional oscillation with two degress of freedom
If the discs of Fig. 2.6(a) are identical of mass 250 kg and radius of gyration 0.5 m coupled by a shaft 800 mm long and 80 mm diameter, calculate the natural frequency of torsional oscillation if the modulus of rigidity of the shaft is 8.0 GN/m².

$$k=\frac{Y}{\theta}=\frac{GJ}{L}=\frac{8\,\mathrm{GN}}{\mathrm{m}^2}\times\frac{\pi}{32}\times\frac{80^4\,\mathrm{mm}^4}{800\,\mathrm{mm}}=40.2\,\mathrm{kN\,m}$$

Since $I_1 = I_2 = I$, then $\frac{I_1 + I_2}{I_1 I_2} = \frac{2}{I}$ and $\omega = \sqrt{\frac{2k}{I}}$

$$\text{i.e. } \omega = \sqrt{\left\{\frac{80.4\ \text{kN}}{250\ \text{kg} \times 0.25\ \text{m}^2}\left[\frac{\text{kg m}}{\text{N s}^2}\right]\right\}} = \sqrt{\frac{12.86 \times 10^2}{\text{s}^2}} = 35.8/\text{s}$$

$$= \frac{35.8\ \text{rad}}{\text{s}}\left[\frac{\text{cycle}}{2\pi\ \text{rad}}\right] = 5.7\ \text{cycle/s} \quad \text{or} \quad 5.7\ \text{Hz}.$$

Considering the case of a two-mass system constrained to move in a straight line as shown diagrammatically in Fig. 2.6(c) vibrating freely (i.e. autonomously), we deduce from the free-body diagrams of the masses that the equations of motion are:

$$m_1\ddot{x}_1 = -k_1 x_1 + k_3(x_2 - x_1)$$

and

$$m_2\ddot{x}_2 = -k_2 x_2 - k_3(x_2 - x_1).$$

Written in matrix form—i.e. in a shorthand way of expressing sets of simultaneous equations (see Appendix B3), we have: $\begin{bmatrix} m_1 & 0 \\ 0 & m_2 \end{bmatrix}\begin{Bmatrix} \ddot{x}_1 \\ \ddot{x}_2 \end{Bmatrix} +$ $\begin{bmatrix} (k_1 + k_3) & -k_3 \\ -k_3 & (k_2 + k_3) \end{bmatrix}\begin{Bmatrix} x_1 \\ x_2 \end{Bmatrix} = \begin{Bmatrix} 0 \\ 0 \end{Bmatrix}$ in which $\begin{bmatrix} m_1 & 0 \\ 0 & m_2 \end{bmatrix}$ is the mass matrix and $\begin{bmatrix} (k_1 + k_3) & -k_3 \\ -k_3 & (k_2 + k_3) \end{bmatrix}$ is the stiffness matrix, the elements in both of

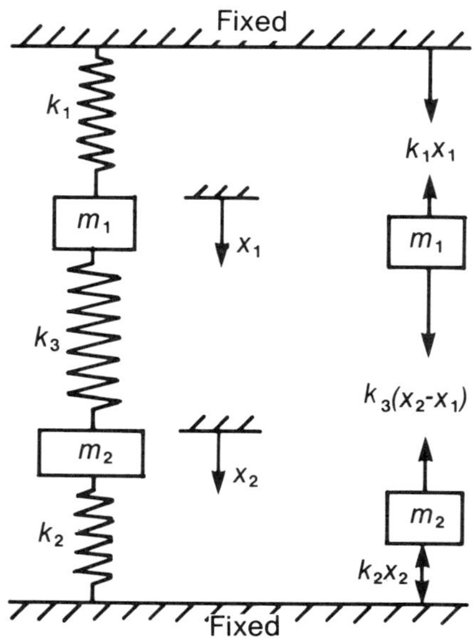

Fig. 2.6(c) Undamped natural oscillators

which are scalar quantities modifying vector quantities represented by the column matrices $\begin{Bmatrix} \ddot{x}_1 \\ \ddot{x}_2 \end{Bmatrix}$ and $\begin{Bmatrix} x_1 \\ x_2 \end{Bmatrix}$.

Assuming a principal mode of motion, both co-ordinates will have harmonic motion of the same frequency, namely $x_1 = X_1 \sin \omega t$ and $x_2 = X_2 \sin \omega t$ which, when substituted in the equations of motion, give the two algebraic simultaneous equations involving the three unknowns X_1, X_2 and ω. Thus:

$$X_1(k_1 + k_3 - m_1\omega^2) - k_3 X_2 = 0$$

$$X_2(k_2 + k_3 - m_2\omega^2) - k_3 X_1 = 0$$

from which

$$\frac{X_2}{X_1} = \frac{k_1 + k_3 - m_1\omega^2}{k_3} = \frac{k_3}{k_2 + k_3 - m_2\omega^2}$$

or

$$m_1 m_2 \omega^4 - \omega^2(m_1(k_2 + k_3) + m_2(k_1 + k_3)) + k_1 k_2 + k_3(k_1 + k_2) = 0.$$

The roots of this frequency equation (a quadratic in ω^2) are the characteristic values (or 'eigen' values) of the system which satisfy the initial assumption that all parts of the system have the same harmonic frequency ω. Hence, the two positive values of ω are natural frequencies for free oscillations in one or other of two principal modes of the system with two degrees of freedom.

For two degrees-of-freedom it is relatively easy to establish a frequency equation and find its roots but, since the order of the frequency equation increases with the number of degrees of freedom, this becomes increasingly difficult for multiple degrees-of-freedom—e.g. for four degrees of freedom, the frequency equation will be a quartic in ω^2, and so on until the establishing of the frequency equation becomes too tedious a task—especially since there are other ways (numerical) of finding the roots of algebraic equations. The roots are, of course, those of a frequency equation which could, if necessary, be arduously determined.

Thus, a numerical solution must be sought for each specific case. The amount of computation involved is prodigious; and for the past three decades, digital computers have been used to assist in this task. However, the advent of the micro-computer has made sufficient computing power much more widely available so that systems with many degrees of freedom can be analysed with appropriate software.

For those having a BBC micro-computer, see Appendix D, (10) and (11).

EXAMPLE 2.6(ii)—illustrative of a preliminary approximate investigation

Start an approximate analysis of the linear vertical and pitching motion of the body of a vehicle.

The most important motions in the complex system of vehicle suspension are the linear vertical motions of the wheels and body, and the pitching or angular motion of the body. The natural frequency of the wheels is usually much greater than that of the main body; and, to simplify the problem, the vertical and angular motions of the body can be dealt with separately.

Referring to Fig. 2.6(d) the co-ordinates y and θ determine the instantaneous linear and angular displacements about the centre of mass, G. The static equilibrium position is shown in Fig. 2.6(e)—the weight of the body ($W = mg$) being supported by the springs exterting forces $R_1 = Wd_2/(d_1 + d_2)$ and $R_2 = Wd_1/(d_1 + d_2)$, whose moments about G balance, i.e. $R_1 d_1 = R_2 d_2$. Also, $R_1 + R_2 = mg$.

The forces on the body when displaced vertically upwards by a distance y are shown on Fig. 2.6(f)—the springs exerting downward forces on the body (i.e. opposite to the positive direction of y); and if the body is then subjected to pitching as well, the forces on the free-body diagram are as shown on Fig. 2.6(g) for a pitching angle θ.

Hence, applying the law $F = m\ddot{y}$, we deduce from Fig. 2.6(g) that $R_1 - k_1(y + d_1\theta) + R_2 - k_2(y - d_2\theta) - mg = m\ddot{y}$ which, since, $R_1 + R_2 = mg$, reduces to $-k_1(y + d_1\theta) - k_2(y - d_2\theta) = m\ddot{y}$.

Similarly, applying the laws $T_G = I_G\ddot{\theta}$ (see Appendix A3) we deduce that $T_G = mk_G^2\ddot{\theta} = F_1 d_1 - F_2 d_2$ or $mk_G^2\ddot{\theta} = R_1 d_1 - k_1 d_1(y + d_1\theta) - R_2 d_2 + k_2 d_2(y - d_2\theta)$ which, since $R_1 d_1 = R_2 d_2$, reduces to $mk_G^2\ddot{\theta} = -k_1 d_1(y + d_1\theta) + k_2 d_2(y - d_2\theta)$. The $F = m\ddot{y}$ and $T_G = I_G\ddot{\theta}$ can be reduced further to $\ddot{y} + py + q\theta = 0$ and $\ddot{\theta} + qy/k_G^2 + \gamma\theta/k_G^2 = 0$ in which $p = (k_1 + k_2)/m$, $q = (k_1 d_1 - k_2 d_2)/m$ and $r = (k_1 d_1^2 + k_2 d^2)/m$.

Thus, if the linear angular motions are harmonic—i.e. $y = Y\cos\omega t$ and $\theta = \Theta\cos\omega t$ we deduce that $\ddot{y} = -Y\omega^2\cos\omega t$ and $\ddot{\theta} = -\Theta\omega^2\cos\omega t$ which, on

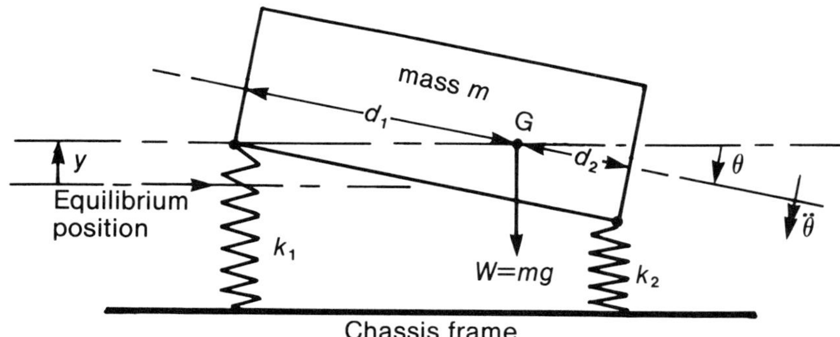

Fig. 2.6(d) Vertical and pitching motion

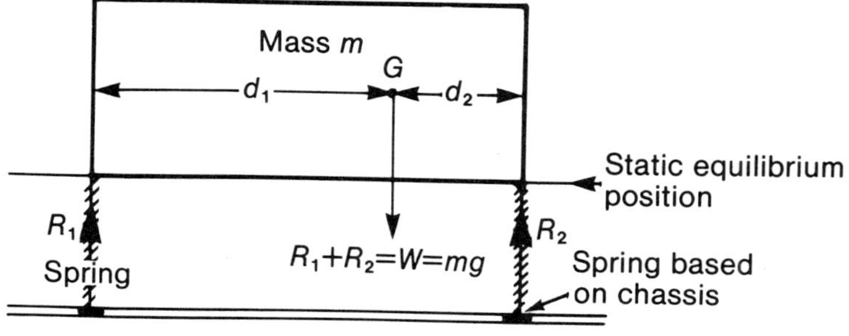

Fig. 2.6(e) Static equilibrium

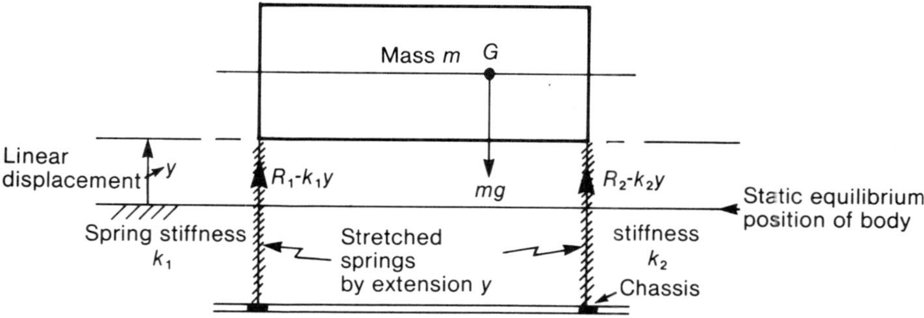

Fig. 2.6(f) Forces on body when displaced vertically upwards

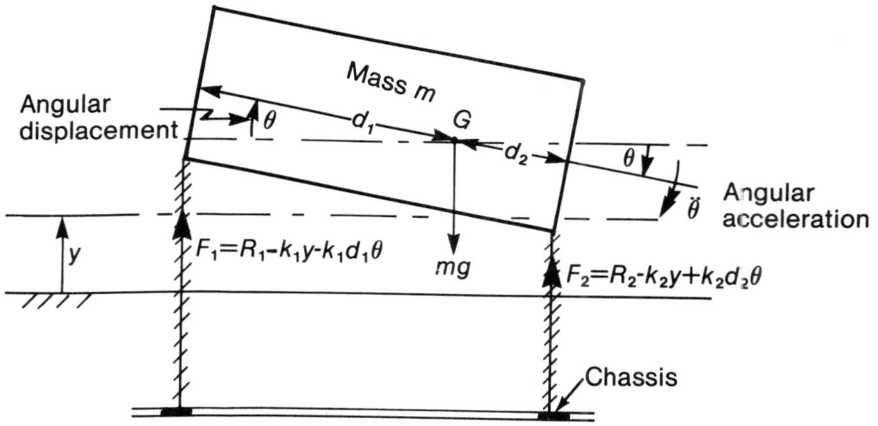

Fig. 2.6(g) Forces when body in vertical and pitching motion

substituting in the last equations, yield

$$(p-\omega^2)Y+q\Theta=0 \quad \text{and} \quad \frac{q}{k_G^2}Y+\left(\frac{r}{k_G^2}-\omega^2\right)\Theta=0$$

from which we get

$$\frac{Y}{\Theta}=-\frac{q}{(p-\omega^2)}=-\frac{(r/k_G^2-\omega^2)}{q/k_G^2}$$

or

$$\omega^4-(p+r/k_G^2)\omega^2+\left(\frac{pr-q^2}{k_G^2}\right)=0$$

The solution of this equation is (see Appendix B2)

$$\omega_{1,2}^2=\tfrac{1}{2}(p+r/k_G^2)\pm\sqrt{\{\tfrac{1}{4}(p-r/k_G^2)^2+(q/k_G)^2\}}$$

from which it follows that if q is small, i.e. $k_1d_1 \doteqdot k_2d_2$ then $\omega_1^2 \doteqdot p$ and $\omega_2^2 \doteqdot r/k_G^2$.

EXAMPLE 2.6(iii)—illustrative of linear (vertical) and angular (pitching) motions of a vehicle

Determine the frequencies and amplitude ratios of a vehicle of mass 1.5 Mg for which k_1 and k_2 are 35 and 40 kN/m; d_1 and d_2 are 1.4 and 1.7m, respectively, and $k_G^2 = 1.5\,\text{m}^2$ referring to Fig. 2.6(d).

$$p = \frac{k_1 + k_2}{m} = \frac{75}{1.5}\frac{\text{N}}{\text{kg m}}\left[\frac{\text{kg m}}{\text{N s}^2}\right] = 50/\text{s}^2$$

$$q = \frac{k_1 d_1 - k_2 d_2}{m} = \left(\frac{49 - 68}{1.5}\right)\frac{\text{N}}{\text{kg}} = -\,12.66\,\text{m/s}^2$$

and

$$\left(\frac{q}{k_G}\right)^2 = 106.7/\text{s}^4$$

$$r = \frac{k_1 d_1^2 + k_2 d_2^2}{m} = 122.7\,\text{m}^2/\text{s}^2$$

and

$$\frac{r}{k_G^2} = 81.7/\text{s}^2$$

Hence,

$$(\omega_{1,2})^2 = \frac{1}{2}\left(p + \frac{r}{k_G^2}\right) \pm \sqrt{\left\{\frac{1}{4}\left(p - \frac{r}{k_G^2}\right)^2 + \left(\frac{q}{k_G}\right)^2\right\}}$$

$$= \frac{1}{2}\left(\frac{131.7}{\text{s}^2}\right) \pm \sqrt{\left\{\frac{1}{4} \times \left(-\frac{31.7}{\text{s}^2}\right)^2 + \frac{106.7}{\text{s}^4}\right\}}$$

$$= \frac{65.8}{\text{s}^2} \pm \sqrt{\frac{131.8}{\text{s}^4}} = \frac{77.3}{\text{s}^2} \quad \text{or} \quad \frac{54.3}{\text{s}^2}$$

i.e.

$$\omega_1 = 8.8/\text{s} = 1.4\ \text{cycle/s}\,;\ \text{and}\ \omega_2 = 7.38/\text{s} = 1.175\ \text{cycle/s}$$

Also,

$$\frac{Y}{\Theta} = -\frac{q}{p - \omega^2} = \frac{12.66\,\text{m/s}^2}{50/\text{s}^2 - 77.3/\text{s}^2} = -\,0.463\frac{\text{m}}{\text{rad}}\left[\frac{\pi\ \text{rad}}{180\ \text{deg}}\right] = 8.1\ \text{mm/deg}$$

EXERCISES ON CHAPTER 2

1. Prove that the frequency of oscillation of a simple pendulum is $\frac{1}{2\pi}\sqrt{\frac{g}{L}}$ for small amplitudes if the mass of the supporting length, L, is negligible.
2. An unknown mass oscillating vertically on the end of a helical spring has a natural frequency of 90/min. When a mass of 0.5 kg is added to the mass the natural frequency is reduced to 78/min. Calculate the unknown mass and the stiffness of the spring.

3. A pulley of mass 30 kg and radius of gyration 200 mm is clamped on a shaft 2 m from one end of a shaft 25 mm diameter and 3 m long. Calculate the natural frequency of torsional vibration neglecting the mass of the shaft which is fixed at both ends. The modulus of rigidity of the material of the shaft is 82.5 GN/m^2.
4. A hydrometer float weighs 0.3 N, and the diameter of the cylindrical section protruding above the surface is 6 mm. Calculate the period of oscillation when the float is allowed to bob up and down in a liquid of relative density 1.2. Density of water is 1 Mg/m^3.
5. A load W is concentrated at a distance c from one end of a simply-supported beam of length L. Neglecting the effect of the mass of the beam show that (i) the stiffness of the beam is given by $k = \frac{W}{d_c} = \frac{3EIL}{c^2(L-c)^2}$, where d_c is the static deflection of the beam at the load position (ii) the periodic time of vibration of the load is $2\pi\sqrt{\frac{d_c}{g}}$, and (iii) the displacement of the load from the equilibrium position is given by $A\cos(\omega_n t + \beta)$.
6. A flywheel of weight W is suspended is a horizontal plane by three wires of length 2 m equally spaced around a circle of radius 250 mm. If the period of small oscillation about a vertical axis through the centre of the wheel is 2 s, find the radius of gyration of the wheel.
7. (a) Using $\frac{M}{EI} = \frac{1}{R} = \frac{d^2y}{dx^2}$, prove that the strain energy or resilience of a bent beam is given by $\frac{1}{2}EI\int_0^L \left(\frac{d^2y}{dx^2}\right)^2 dx$, where y is the deflection at a point x in the span of length L.

   (b) Using $\frac{T}{J} = \frac{G\theta}{L}$, show that the strain energy of a circular shaft in torsion is given by $\frac{1}{2}\frac{GJ}{L}\theta^2 = \frac{1}{2}T\theta$.

   (c) Using $q = C\theta$ for shear stress, show that the strain of a body in pure shear is given by $\frac{1}{2}q\theta \times$ volume, where θ is the shear strain.
8. A bar of length L (assumed massless) is pivoted at one end and carries a mass M at the other end. It is supported by an inextensible cord of length h at a distance a from the pivot, and moves with s.h.m. Calculate the natural angular frequency for small oscillations using $T_{max} = V_{max}$.
9. A reed-type of frequency-measuring instrument consists of small cantilever beams with masses attached to the ends. Using $T_{max} = V_{max}$, estimate the mass which must be placed at the end of a reed made of steel 1 mm thick, 6 mm wide, 90 mm long and of $E = 200$ GN/m^2 to measure a frequency of vibration 20 cycles/s. Derive any formula used, and state any assumptions made.
10. A mass of 4.5 kg attached to the lower end of a vertical helical spring whose upper end is fixed, oscillates with a natural period of 0.5 s. Estimate the stiffness of the spring and the natural period when a mass weighing 235N is attached to the mid-position of the same vertical spring with both ends fixed.
11. Deduce a formula for the frequency of the natural torsional oscillation of a system composed of two identical wheels of moment of inertia I keyed on to the ends of a uniform shaft of length L, modulus of rigidity G and second moment of area J about the axis.

12. Find the natural frequency of a double pendulum with equal masses on equal lengths, assuming small oscillations. Also, determine the ratio of the amplitudes of the two modes associated with the two degrees of freedom.
13. A cylinder of total mass m and cross-sectional area A is weighted at the base so as to float vertically in liquid of density ρ. Neglecting accompanying motion of fluid, estimate the periodic time of the oscillating cylinder when disturbed vertically.

3

Damped natural vibrations

3.1 TYPES OF DAMPING

Although in practice it is rare to find the forms of damping occurring exclusive of one another, mathematical expediency usually requires such an assumption. Damping is the reason for decay or reduction in amplitude of vibrations, and is associated with dissipation of energy usually caused by one, or a combination, of:

(i) viscous damping encountered by bodies moving in fluids (liquid or gaseous)—the force of resistance (cv) being assumed proportional to the relative velocity (v) which enables the equation of motion to be kept linear, and in which c is the damping coefficient,

(ii) Coulomb damping caused by the relative motion of dry surfaces—the force (μR) being proportional to the normal component (R) of the force between the two surfaces (i.e. Coulomb's name is associated with damping which does not depend on relative velocity between surfaces)—the proportionality constant (μ) is the coefficient of friction,

(iii) solid (or hysteretic) damping caused by internal friction or hysteresis when a solid is deformed, e.g. the vibration causing sound from a bell dies away because of hysteretic damping, and

(iv) electrical damping caused by the cutting or crossing of magnetic fields.

In a typical mechanical structure or mechanism, damping is not localized (or 'lumped') and viscous but is usually distributed and mostly non-linear, i.e. not simply proportional to velocity. Thus, the inevitable assumptions which have to be made (see Example 3.8(ii)) in order to make analysis possible necessarily imply that in any practical vibration system, predictions regarding its behaviour can only be approximate. Fact-finding by means of instruments and measurements on actual systems enables assessments to be made on the reliability of the assumptions made for analytical and prediction purposes.

EXAMPLE 3.1(i)—illustrative of dissipation of energy by viscous damping
The instantaneous rate of dissipation of energy is frictional force × velocity, i.e. $(c\dot{x}) \times \dot{x} = c\dot{x}^2$. Hence, the energy dissipation for any particular cycle is $\int_0^{\tau_d} c\dot{x}^2 \mathrm{d}t$, where $\tau_d = 2\pi/\omega_d$. For steady sinusoidal motion $x = X \sin \omega t$, where $\omega = \omega_n \sqrt{(-\zeta^2)} = \omega_d$ (as in Section 3.3). Thus, the integral becomes (see Appendix B5)

$$X^2\omega c \int_0^{2\pi/\omega} \cos^2 \omega t \,\mathrm{d}(\omega t) = \frac{X^2\omega c}{2} \int_0^{2\pi/\omega} (1 + \cos 2\omega t)\mathrm{d}(\omega t) = X^2\pi\omega c.$$

This value decreases in successive cycles because amplitude, X, decreases with time (see Fig. 3.5(a)).

3.2 DAMPED NATURAL OSCILLATING SYSTEMS

The essential features are mass, elastic stiffness, viscous damping and one degree of freedom. Fig. 3.2(a) shows the simple system of a support (assumed rigid) to which is attached a helical spring suspending a mass assumed to be in vertical linear oscillation (i.e. having one degree of freedom) resisted by viscous damping. Thus, the force of resistance is assumed to be proportional to the velocity of the dashpot-piston relative to the fixed guides; and the mass of the system is assumed localized or 'lumped' as m.

Natural oscillation implies that no external force is applied other than that of gravity which acts all the time but is balanced by the spring force, and establishes the system in state of static equilibrium prior to oscillatory motion (as in Section 2.2). Many actual vibrating systems can be reduced or 'idealized' to this simple damped system shown diagrammatically in Fig.

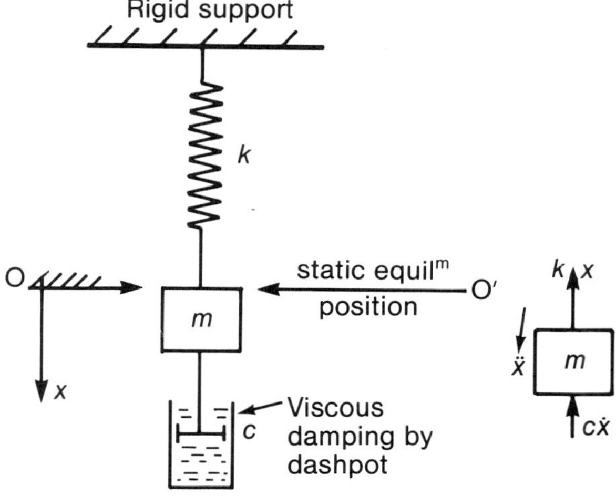

Fig. 3.2(a) Damped natural oscillating system

3.2(a); and predictions from the analysis of which are useful pointers as to what is happening in more sophisticated systems of actual machinery which may not look anything like the diagrammatic representation in Fig. 3.2(a) but which have the same essential features.

If the mass moves vertically downwards (i.e. in the positive direction) through a small displacement x from the position of static equilibrium (oo′), the force exerted on the mass m by the spring is—kx (i.e. upwards), where k is the spring stiffness (force/displacement). Also, when the displacement is x, the positive (downward) velocity is $\frac{dx}{dt}$ or $\dot{x}$, and the force of viscous resistance which acts on the mass is $-c\frac{dx}{dt}$ or $-c\dot{x}$ where c is the damping coefficient (force of resistance/velocity of the mass). The downward acceleration of the mass is $\frac{d^2x}{dt^2}$ or $\ddot{x}$. Hence, as deduced from the free-body diagram on Fig. 3.2(a) the equation of motion (recognising the fact $mg = kd$ as in Section 2.2) is

$$-c\dot{x} - kx = m\ddot{x} \quad \text{or} \quad \ddot{x} + \frac{c}{m}\dot{x} + \frac{k}{m} = 0 \tag{1}$$

The solution of this equation must be a function that retains its form when differentiated with respect to time. Thus, assuming $x = X\varepsilon^{\omega t}$, we have $\omega^2 + \frac{c}{m}\omega + \frac{k}{m} = 0$, where ω has to be determined. This equation (see Appendix B2) has roots $\omega_{1,2} = -\frac{c}{2m} \pm \sqrt{\left\{\left(\frac{c}{2m}\right)^2 - \frac{k}{m}\right\}}$ or, written as complex numbers (see Appendix B9), $\omega_{1,2} = -\frac{c}{2m} \pm j\omega_d$, where $\omega_d = \sqrt{\left\{\frac{k}{m} - \left(\frac{c}{2m}\right)^2\right\}}$, and the general solution may be written (see Appendix B12) $x = A\varepsilon^{\omega_1 t} + B\varepsilon^{\omega_2 t}$ except when the radical is zero which defines the 'critical damping' as follows. The expression for ω changes from real to imaginary when $\left(\frac{c_c}{2m}\right)^2 = \frac{k}{m}$, and the value of the damping coefficient at which this transition occurs is designated 'critical damping' $c_c = 2\sqrt{(km)}$ for which the general solution is $x = (A + Bt)\varepsilon^{\omega_n t}$, where $\omega_n = \sqrt{(k/m)} = \frac{c_c}{2m}$ and the arbitrary constants, A and B, depend on how the motion is started. We should note that, as previously, ω_n is the angular frequency of natural oscillation without damping (Fig. 2.2(a)), i.e. ω_n is the undamped natural frequency (see Chapt. 2).

The amount of damping in any system can be conveniently specified or expressed by the ratio of the damping coefficients c/c_c, i.e. non-dimensionally as the 'damping factor', namely, $\zeta = \dfrac{c}{c_c} = \dfrac{c}{2\sqrt{(mk)}} = \dfrac{c}{2m\omega_n}$. Thus, ζ denotes the ratio of the actual damping in a system to critical damping, i.e. is a dimensionless index of damping in a single degree-of-freedom system. The reason for introducing ζ is that its value determines the character of the motion of the system since the value of this damping factor ζ depends on the damping coefficient, c, the mass, m, and stiffness, k, of the spring or elastic system. Thus, using the critical damping coefficient, c_c, as a reference, it follows that the two roots may be expressed as

$$\omega_{1,2} = \frac{c_c}{2m}\left\{-\frac{c}{c_c} \pm \sqrt{\left\{\left(\frac{c}{c_c}\right)^2 - 1\right\}}\right\} = \omega_n\{-\zeta \pm \sqrt{(\zeta^2 - 1)}\}$$

and the equation of motion (1) may be re-written as

$$\ddot{x} + 2\zeta\omega_n\dot{x} + \omega_n^2 x = 0 \tag{2}$$

in which $2\zeta\omega_n = c/m$ and $\omega_n^2 = k/m$. Alternatively, $[\mathrm{D}^2 + 2\zeta\omega_n\mathrm{D} + \omega_n^2]x = 0$.

i.e.
$$[\mathrm{D} + \omega_n(\zeta + \sqrt{(\zeta^2 - 1)})][\mathrm{D} + \omega_n(\zeta - \sqrt{(\zeta^2 - 1)})]x = 0 \tag{2a}$$

from which it follows that the behaviour of damped natural systems depends on whether $\sqrt{(\zeta^2 - 1)}$ is imaginary (i.e. $\zeta = 0$ or $0 < \zeta < 1$), zero (i.e. $\zeta = 1$, critical damping) or real (i.e. $\zeta > 1$).

Of these four cases, $\zeta = 0$—i.e. undamped oscillation, have been dealt with in Section 2.3, and the others are detailed in Sections: 3.3, for light damping $0 < \zeta < 1$; 3.4 for critical damping, or dead-beat damping, $\zeta = 1$; 3.5 for heavy damping $\zeta > 1$—the two latter prevent any oscillation, as we shall see (also, see, Appendix B12).

3.3 DAMPED NATURAL MOTION WITH LIGHT DAMPING ($0 < \zeta < 1$)

This condition is the commonest one in practice, namely, $0 < \dfrac{c}{2\sqrt{(mk)}} < 1$, where the damping coefficient, c, is positive. The fundamental equation (2) in Section 3.2 applies, namely, $\ddot{x} + 2\zeta\dot{x} + \omega_n^2 x = 0$, the solution of which is often expressed in the form (see Appendix B12):

$$x = \varepsilon^{-\zeta\omega_n t}\{C_1 \cos(\omega_n\sqrt{(1 - \zeta^2)})t + C_2 \sin(\omega_n\sqrt{(1 - \zeta^2)})t\}$$

where C_1 and C_2 are constants; but this equation is rather cumbersome compared with the easier and more useful form e.g. for purposes of diagrammatic representation of x, $\dot{x}$ and $\ddot{x}$ or x, v and a on a polar or 'phase diagram' such as Fig. 3.3(d), namely,

$$\boldsymbol{x = C\varepsilon^{-\zeta\omega_n t}\cos(\omega_d t + \beta) = X\cos(\omega_d t + \beta)} \tag{1}$$

where C is a constant (whose value will latter be seen to depend on initial conditions) and β is the initial phase of the displacement of vector OB when $t = 0$, and

$$\boldsymbol{\omega_d = \omega_n\sqrt{(1 - \zeta^2)}} \tag{1a}$$

The constants C_1, C_2, C and β are inter-related in accordance with Fig. 3.3(a). The cosine term in equation (1) shows that the motion is oscillatory with a periodic time $\tau_d = 2\pi/\omega_d$, but the exponential shows that the amplitude ($X = C\varepsilon^{-\zeta\omega_n t}$) decreases with time as shown in Fig. 3.3(b).

Differentiating equation (1) gives the instantaneous velocity

$$\dot{x} = v = \omega_n C\varepsilon^{-\zeta\omega_n t}\cos\left(\omega_d t + \beta + \frac{\pi}{2} + \psi\right)$$

or

$$v = V\cos\left(\omega_d t + \beta + \frac{\pi}{2} + \psi\right) \tag{2}$$

where $V = \omega_n X = C\omega_n\varepsilon^{-\zeta\omega_n t}$, and $\psi = \sin^{-1}\zeta$ and $\cos\psi = \sqrt{(1 - \zeta^2)}$ in accordance with Fig. 3.3(c). Also, differentiating equation (2) gives the instantaneous acceleration $\ddot{x} = a = \omega_n^2 C\varepsilon^{-\zeta\omega_n t}\cos(\omega_d t + \beta + \pi + 2\psi)$ or

$$\boldsymbol{a = A\cos(\omega_d t + \beta + \pi + 2\psi)} \tag{3}$$

where $A = \omega_n^2 X = C\omega_n^2\varepsilon^{-\zeta\omega_n t}$

The similarity and comparison of equations 1, 2 and 3 (Section 2.3) for undamped natural (free or autonomous) oscillations and 1, 2 and 3 for damped natural oscillations shows that ω_d plays the same part in the latter set as does ω_n in the former, and, since $\zeta\left(\text{i.e. } \dfrac{c}{2\sqrt{(ms)}}\right)$ is a pure number, both

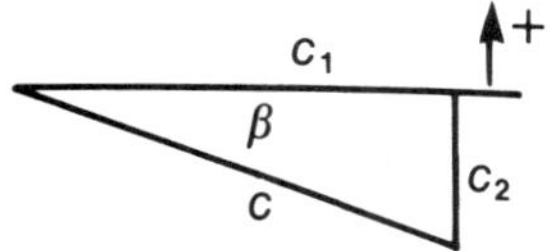

Fig. 3.3(a) Interrelation of constants

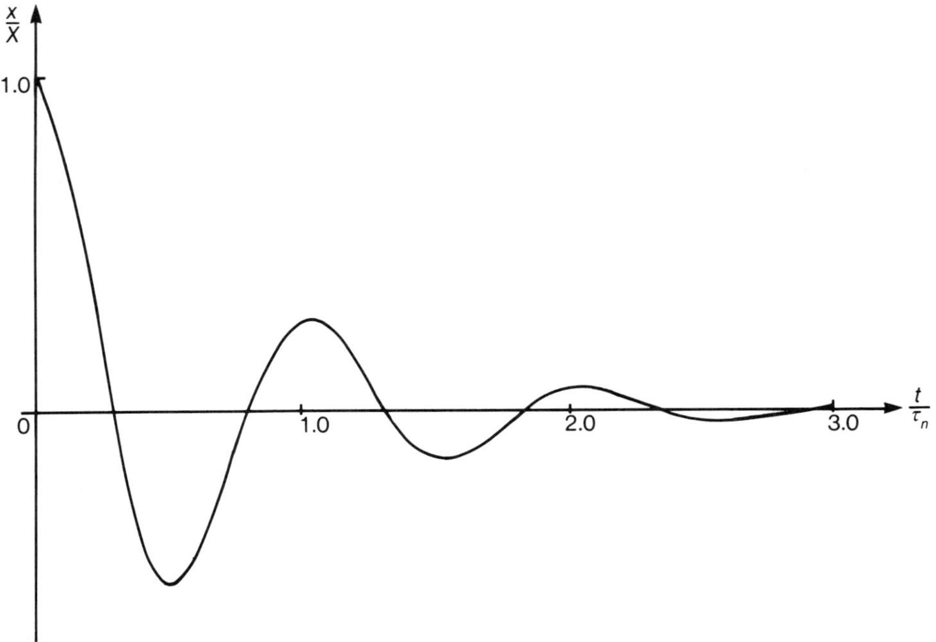

Fig. 3.3(b) Decay curve for damping factor $\zeta = 0.2$

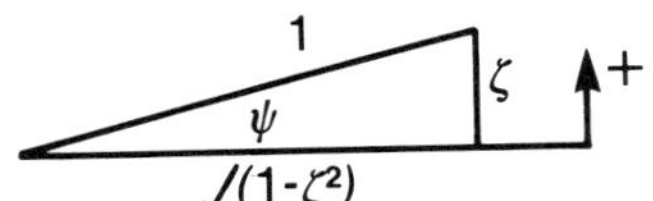

Fig. 3.3(c) Interrelation of constants

have the same dimension $\left[\frac{1}{T}\right]$. Thus, ω_d can be regarded as the angular frequency of a damped oscillation, and is nearly equal to ω_n when the damping factor ζ is small. Also, the phase difference $\left(\frac{\pi}{2} + \psi\right)$ between the vectors representing displacement (OB) and velocity (OC) is the same as that between those representing velocity (OC) and acceleration (OD) shown in Fig. 3.3(d) as compared with a phase difference of $\pi/2$ in the case of undamped natural oscillations shown in Fig. 2.3(a). Thus, in the same way as equations 1, 2 and 3 in Section 2.3 have a simple graphical representation or interpretation on Fig. 2.3(a), so have equations 1, 2 and 3 in this Section on Fig. 3.3(d). On the latter, OB is drawn at angle β (counterclockwise if positive so that when $t = 0$, $x = X \cos \beta$) from the zero time-line OA and of length $X = C\varepsilon^{-\zeta\omega_n t}$

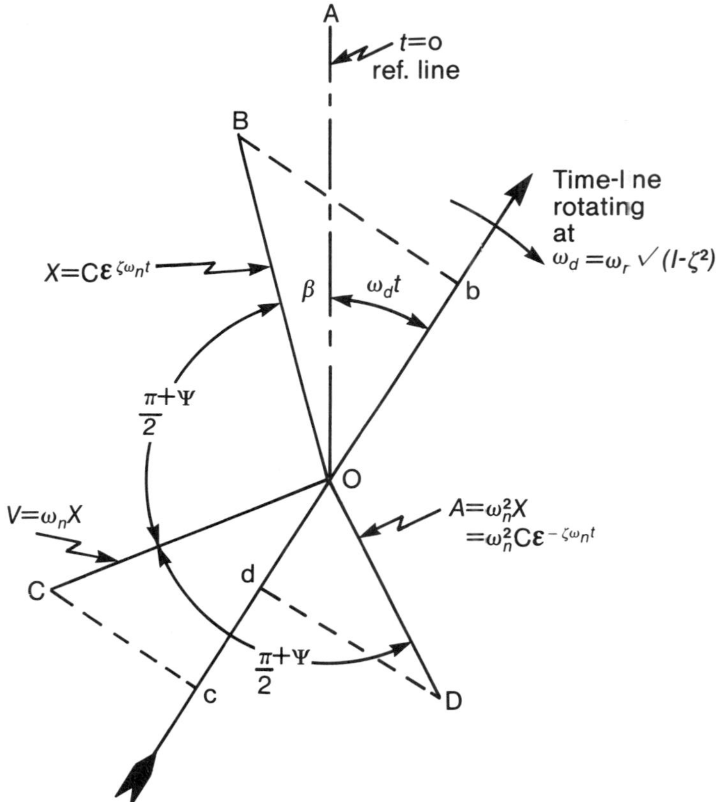

Fig. 3.3(d) Phase diagram for damped natural oscillations

representing the maximum value of the amplitude or displacement of the mass m shown in Fig. 3.2(a). Similarly, OC is drawn at angle $(\pi/2+\psi)$ counterclockwise from OB and of length $V=\omega_n X$ representing the maximum value of the velocity of the mass m; and OD is drawn at an angle $(\pi/2+\psi)$ counterclockwise from OC and of length $A=\omega_n^2 X$ representing the maximum value of the acceleration of mass m.

The time-line rotates clockwise at an angular rate $\omega_d=\omega_n\sqrt{(1-\zeta^2)}=\omega_n\cos\psi$ which is slower than that (ω_n) for an undamped oscillation; and the time for one revolution of the time-line is, therefore, slightly longer than when there is no damping. The instantaneous values of displacement, velocity and acceleration of the oscillating mass (Fig. 3.2(a)) at any time t are the projections on the time-line, namely, ob, oc and od, respectively; but, because of the damping, the lengths of vectors OB, OC and OD shrink with time since each depends on the exponential term $\varepsilon^{-\zeta\omega_n t}$.

In the vectors themselves rotated instead of the time-line, the tips of the three vectors would trace-out logarithmic spirals finishing at O this signifying that the oscillation had been damped out. Alternatively, the time-dependent behaviour of the system can be illustrated if the instantaneous values of displacement (see equation 1) are plotted against time. We get the 'decay curves' for natural damped oscillations as shown in Fig. 3.3(e) by non-dimensional plotting, where curves for values of the damping factor $\zeta = 0.1$ and 0.2 have been superimposed on the case for undamped ($\zeta = 0$) motion which is, of course, simple harmonic.

Equation (1), written non-dimensionally, is

$$\frac{x}{X_o} = \varepsilon^{-\zeta\omega_n t} \cos\left(2\pi\left(\frac{t}{\tau_d}\right) + \beta\right) \text{ if at } t = 0,\ x = X_0 \text{ and } \dot{x} = 0,$$

or

$$\frac{x}{X_o} = \varepsilon^{-2\pi\zeta(t/\tau_n)} \cos\left(2\pi\left(\frac{t}{\tau_n}\sqrt{(1-\zeta^2)}\right) + \beta\right),$$

where $\beta = \tan^{-1}\left(\frac{\zeta}{\sqrt{(1-\zeta^2)}}\right)$ when $t = 0$, $\dot{x} = 0$ as determined from $\dot{x} = C\omega_n\{-\zeta\cos\beta - \sqrt{(1-\zeta^2)}\sin\beta\} = 0$, (also, see Example 3.5(i)).

Readers with access to a micro-computer with high-resolution graphics could usefully plot equation (1) for various initial conditions and values for $\zeta(0 \leqslant \zeta \leqslant 1)$. This would reinforce and extend the information shown on Fig. 3.3(e).

As regards the forces involved, we recollect that kx, $c\dot{x}$ (or cv) and $m\ddot{x}$ (or ma) are the forces of the spring, viscous resistance and inertia force of the mass, respectively. Hence, the maximum values of these forces

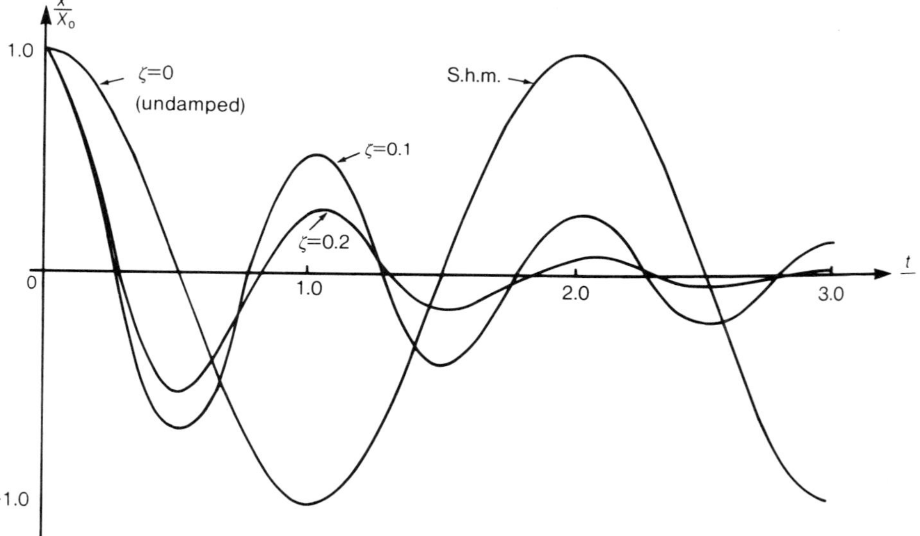

Fig. 3.3(e) Decay curves for damped natural oscillations

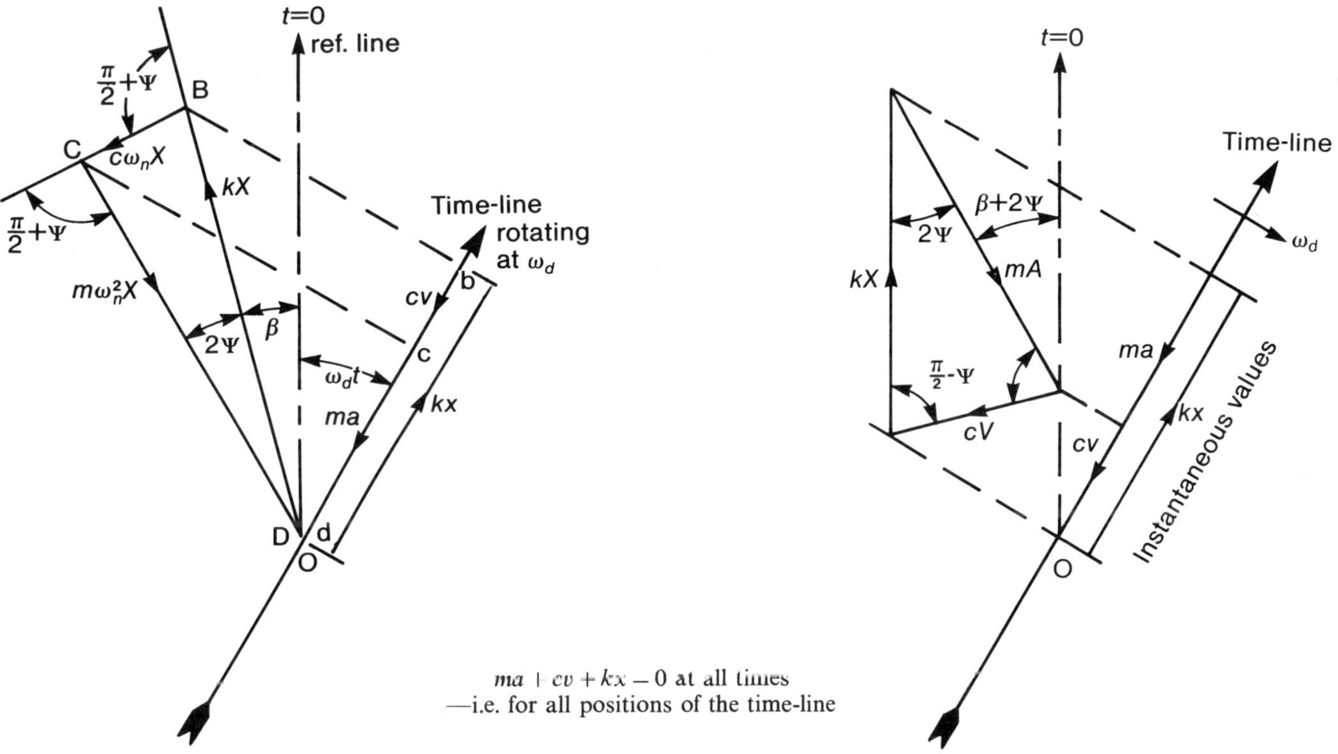

Fig. 3.3(f and g) Phase diagrams of forces of damped natural oscillations

are obtained when the vectors X, V and A are multiplied by their appropriate coefficients k, c and m, respectively.

Since, by definition of natural oscillation, no external force (other than gravitational) is applied to an m, k, c system, the instantaneous forces will have zero resultant at all times during the cycle in accordance with the original equation of forces, namely, $m\ddot{x} + c\dot{x} + kx = 0 = ma + cv + kx$. The instantaneous values of the forces are shown on Fig. 3.3(f) as ob, bc and cd—i.e. as projections of kX, cV and mA, respectively, on the rotating time-line.

Hence, the maximum values of the forces kX, cV and mA must form a closed vector diagram of forces as shown on Fig. 3.3(f) or alternatively on Fig. 3.3(g) both of which, of course, shrink and vanish with time in accordance with equations (1), (2) and (3) because of the exponential factor ($\varepsilon^{-\zeta\omega_n t}$) which the vectors contain.

EXAMPLE 3.3(i)—illustrative of coefficient of viscosity determined from periodic times

As a means of finding the dynamic viscosity (η) of a liquid, a thin plate of total surface area $2A$ and mass m is attached to the end of a helical spring and allowed to oscillate vertically in the fluid. If τ_1 is the natural period of undamped oscillation (i.e. when oscillating in air) and τ_2 is the damped period when the plate oscillates immersed in the liquid, deduce a formula from which η can be calculated.

Referring to Fig. 3.3(h), and applying Newton's second law of motion, we get: $m\ddot{x} = W - c\dot{x} - k(d + x)$, where d is static extension of spring, or $\ddot{x} + 2k\omega_n\dot{x} + \omega_n x = 0$, where $W = kd$, $2\zeta\omega_n = c/m$, $\omega_n = \sqrt{\left(\frac{k}{m}\right)} = \frac{2\pi}{\tau_1}$. Also, by definition, $\eta = \frac{F}{2A\dot{x}}$ and $c = \frac{F}{\dot{x}} = 2A\eta$. Hence, $\frac{c}{2} = A\eta = m\zeta\omega_n$ and $\zeta = \frac{A\eta}{m\omega_n}$. Also, since $\omega_d = \omega_n\sqrt{(1 - \zeta^2)}$ and $\tau_1\omega_n = 2\pi = \tau_2\omega_d$ then

$$\frac{\omega_d}{\omega_n} = \frac{\tau_1}{\tau_2}\sqrt{\left\{1 - \left(\frac{A\eta}{m\omega_n}\right)^2\right\}} \quad \text{and} \quad \left(\frac{A\eta}{m\omega_n}\right)^2 = 1 - \left(\frac{\tau_1}{\tau_2}\right)^2 = \frac{\tau_2^2 - \tau_1^2}{\tau_1^2}.$$

Hence,

$$\eta = \frac{2\pi}{\tau_1\tau_2}\frac{m}{A}\sqrt{(\tau_2^2 - \tau_1^2)}.$$

EXAMPLE 3.3(ii)—illustrative of experimental estimation of damping coefficient

A torsional system composed of a disc-wheel of moment of inertia I on the end of a shaft of torsional (T/θ) stiffness k has a natural cyclic frequency f_1 in air and f_2 when immersed in oil. Relate these in a formula to give the damping coefficient (c).

Torque $= I\ddot{\theta} = -k\theta - c\dot{\theta}$ or $\ddot{\theta} + \frac{c}{I}\dot{\theta} + \frac{k}{I}\theta = 0$, where $k = T/\theta$. Alternatively, using

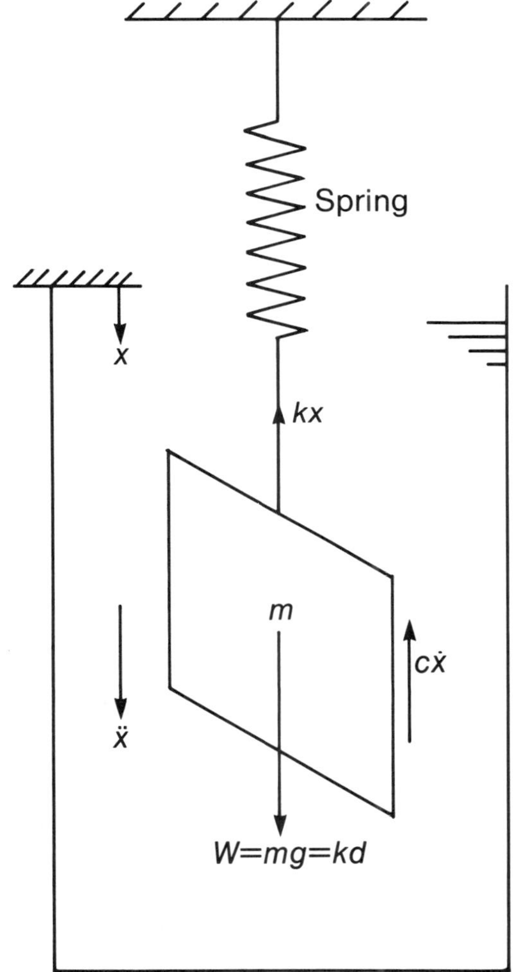

Fig. 3.3(h) Oscillating plate in liquid

$\omega^{2n} = k/l$ and $2\zeta\omega_n = c/I$, i.e. $\zeta = \dfrac{c}{2\sqrt{(kI)}}$, we have $[D^2 + 2\zeta\omega_n D + \omega_n^2]\theta = 0$ as the differential equation for torsional oscillations. The solution of the equation (see Appendix B12) may be written $\theta = \Theta\cos(\omega_d t + \beta)$ where Θ represents the maximum angle of twist of the shaft, and $\omega_d = \omega_n\sqrt{(1-\zeta^2)}$. Hence, $\dfrac{\omega_d}{\omega_n} = \sqrt{(1-\zeta^2)} = \dfrac{f_2}{f_1}$, where $f_1 = \dfrac{\omega_n}{2\pi} = \dfrac{1}{2\pi}\sqrt{(k/I)}$. It follows that $\zeta^2 = 1 - \left(\dfrac{f_2}{f_1}\right)^2 = \dfrac{c}{4kI}$ and, hence $c^2 = \dfrac{4kI}{f_1^2}(f_1^2 - f_2^2) = 16\pi^2 I^2(f_1^2 - f_2^2)$ i.e. $c = 4\pi I\sqrt{(f_1^2 - f_2^2)}$.

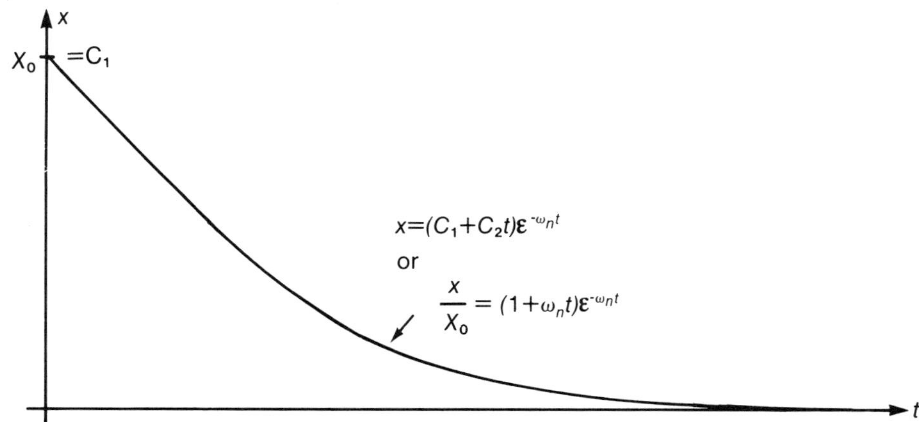

Fig. 3.4(a) Dead-beat motion

3.4 DAMPED NATURAL MOTION WITH CRITICAL DAMPING ($\zeta = 1$)

The case when $\zeta = 1$ is called 'critical damping' because it represents the transition between oscillatory and non-oscillatory (or dead-beat) motion, e.g. a weigh-scale pointer.

When $\zeta = 1$ the general equation (2) in Section 3.2 becomes $\ddot{x} + 2\omega_n\dot{x} + \omega_n^2 x = 0 = [\mathrm{D} + \omega_n]^2 x$, but its *general* solution is *not* $x = C_1\varepsilon^{-\omega_n t} + C_2\varepsilon^{-\omega_n t} = C\varepsilon^{-\omega_n t}$ because this contains only one arbitrary constant instead of two which are required in the solution of a second order differential equation (see Appendix B11). However, $x = C\varepsilon^{-\omega_n t}$ is a solution and, if we let $x = z\varepsilon^{-\omega_n t}$ i.e. making z the new independent variable, we deduce (see Appendix B12) that $[\mathrm{D}^2 + 2\omega_n \mathrm{D} + \omega_n^2]z\varepsilon^{-\omega_n t} = 0 = \varepsilon^{-\omega_n t}[\mathrm{D}^2]z$. Hence, $[\mathrm{D}^2]z = 0$ and $z = (C_1 + C_2 t)$, and the general solution is $x = (C_1 + C_2 t)\varepsilon^{-\omega_n t}$. In this, $\varepsilon^{-\omega_n t}$ never vanishes as is shown on Fig. 3.4(a), i.e. the mass just fails to overshoot as it returns to the equilibrium or mid-position (also, see Fig. 3.6(a)). In particular, if the maximum value of x occurs when $t = 0$, then $X_0 = C_1$ and, since the velocity is zero when $t = 0$, $C_2 - \omega_n C_1 = 0$ or $\omega_n = C_2/C_1$. Hence, when $\zeta = 1$, motion is critically damped and $\dfrac{x}{X_0} = (1 + \omega_n t)\varepsilon^{-\omega_n t}$ as in Example 3.5(i).

EXAMPLE 3.4(i)—illustrative of the design of a dead-beat manometer

A U tube manometer of uniform bore is required to read the difference in air pressure between two compartments. Find the total length of a water column necessary to a frequency of 1 cycle/s if the motion is undamped. Also, estimate the diameter of the tube which will render the motion dead-beat. $\dfrac{\mathrm{d}P}{\mathrm{d}L} = 8\eta\dfrac{v}{r^2}$ for laminar flow. For water $\rho = 1\ \mathrm{Mg/m^3}$ and $\eta = 1\ \mathrm{g/sm}$.

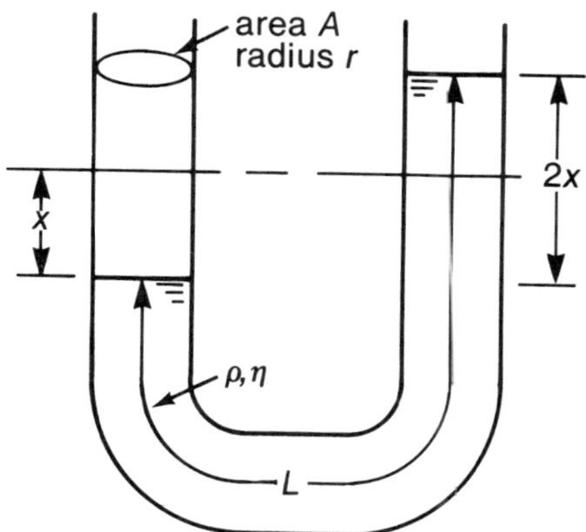

Fig. 3.4(b) Motion of liquid

Referring to Fig. 3.4(b), for a small displacement x the restoring force is $2xA\rho g = kx$, and $\omega_n = \sqrt{(k/m)} = \sqrt{\left(\frac{2A\rho g}{\rho AL}\right)} = \sqrt{\left(\frac{2g}{L}\right)}$—(see, also, Example 2.4(i)). Since $\tau_n = \frac{2\pi}{\omega_n} = 1\text{s}$, then $\left(\frac{2\pi}{\text{s}}\right)^2 = \frac{2g}{L}$ and $L = \frac{g}{2\pi^2}\text{s}^2 = \frac{9811\,\text{mm}}{2\pi^2} = 497\,\text{mm}$.

If the motion is to be dead-beat (i.e. $\zeta = 1$) then

$$\zeta = \frac{c}{2\sqrt{(mk)}} = \frac{c}{2m}\sqrt{\left(\frac{m}{k}\right)} = \frac{c}{2m\omega_n} = 1, \text{ i.e. } c = 2m\omega_n.$$

Also, $\frac{\text{d}P}{\text{d}L} = 8\eta\frac{v}{r^2}$, where $\frac{\text{d}P}{\text{d}L}$ is the change in pressure along the elementary length $\text{d}L$ of the tube. Hence, the total pressure drop in the U tube is $P = 8\eta\frac{vL}{r^2}$, and the force $F = \pi r^2 P = 8\pi\eta vL$, and the damping coefficient $c = \frac{F}{v} = 8\pi\eta L$. Thus, $8\pi\eta L = 2m\omega_n = 2\rho AL\omega_n$ and $A = \frac{4\pi\eta}{\omega_n\rho} = \frac{\pi}{4}d^2$ from which we deduce

$$d = 4\sqrt{\left(\frac{\eta}{\rho\omega_n}\right)} = 4\sqrt{\left\{\frac{0.001\,\text{m}^2}{1000\;\text{s}} \times \frac{\text{s}}{2\pi}\left[\frac{10^6\text{mm}^2}{\text{m}^2}\right]\right\}}$$
$$= 4\sqrt{\left(\frac{\text{mm}^2}{2\pi}\right)} = 1.6\,\text{mm}$$

EXAMPLE 3.4(ii)—illustrative of the design of a recoil-return system
A large gun is designed so that when fired the barrel recoils against a spring. At the end of the recoil a dashpot is engaged and allows the barrel to return to its initial position in minimum time without oscillations. Determine the necessary spring constant (k) and

dashpot coefficient (c) for a barrel of mass 725 kg if the initial recoil velocity at the instant of firing is 25 m/s and the recoil distance is 1.5m.

Since the dashpot is not operative during recoil, the initial kinetic energy of the barrel is equal to the work done on the spring, i.e. $\frac{1}{2}mv^2 = \frac{1}{2}kx^2$.

Thus,

$$k = m\left(\frac{v}{x}\right)^2 = 725\text{kg}\left(\frac{25}{1.5\text{s}}\right)^2\left[\frac{\text{Ns}^2}{\text{kg m}}\right] = 201\ \text{kN/m}$$

For the least time of return of the barrel, critical damping is required, i.e. $\zeta = 1 = \dfrac{c}{2\sqrt{(mk)}}$.

Thus

$$c = c_c = 2\sqrt{(mk)} = 2\sqrt{\left\{725\ \text{kg} \times 201\frac{\text{kN}}{\text{m}}\left[\frac{\text{Ns}^2}{\text{kg m}}\right]\right\}}$$

$$= 24\ \text{kN m/s or kN s/m}$$

3.5 DAMPED NATURAL MOTION WITH HEAVY DAMPING ($\zeta > 1$)

In this case the general equation (2) in Section 3.2 has two unequal roots, namely, $\omega_{1,2} = -\omega_n\{\zeta \pm \sqrt{(\zeta^2 - 1)}\}$ which are both negative since the radical $\sqrt{(\zeta^2 - 1)}$ is real and less than ζ.

Hence, $x = C_1\varepsilon^{\omega_1 t} + C_2\varepsilon^{\omega_2 t}$ is the sum of two decaying exponentials with initial value $C_1 + C_2$. The motion is not periodic, and no vibration occurs; instead, the motion causes a creep back to the equilibrium position. This type of motion is sometimes called 'aperiodic' but is of little practical interest—being overdamped.

EXAMPLE 3.5(i)—illustrative of the effect of damping factor on oscillations of the pointer of an instrument

An instrument (which may be considered equivalent to a mass on a massless spring with viscous damping) has 100 divisions on either side of a central zero. If the pointer is held steady at + 100 and suddenly released, plot curves non-dimensionally showing the values of readings against time for values of the damping factor $\zeta = 0.1$, 0.2, 0.6, 1.0, 2.0.

Choosing the non-dimensional co-ordinates x/X_0, where X_0 is the maximum amplitude, and $\dfrac{t}{\tau_n}$ where $\tau_n = 2\pi/\omega_n$, we may state (also see Appendix B12) for this spring-control system with viscous damping that when:

(a) $\mathbf{0 < \zeta < 1}$, $x = C\varepsilon^{-\zeta\omega_n t}\cos(\omega_d t + \beta) = X_0\cos(\omega_d t + \beta)$ from Section 3.3. Thus, when $t = 0$, $x = 100$ divisions $= X_0\cos\beta = C\cos\beta$ and $C = X_0$. Also, since

$$\omega_d = \omega_n\sqrt{(1 - \zeta^2)},\ \text{when}\ t = 0,$$

$$\dot{x} = 0 = C\omega_n\{-\zeta\cos\beta - \sqrt{(1 - \zeta^2)}\sin\beta\}$$

and hence $\tan\beta = -\zeta/\sqrt{(1-\zeta^2)}$ and $\cos\beta = \sqrt{(1-\zeta^2)}$. Therefore, the non-dimensional equation may be written:

$$x/X_0 = \varepsilon^{-\zeta 2\pi t/\tau_n} \cos(\sqrt{(1-\zeta^2)}2\pi t/\tau_n + \beta).$$

Hence, for $\zeta = 0.1$, $\beta = \tan^{-1}(-0.1/\sqrt{0.99}) = -5°56'$ and

$$x/X_0 = \varepsilon^{-0.628t/\tau_n} \cos(358° t/\tau_n - 5°56')$$

which yields the following table of results.

t/τ_n	0	0.2	0.4	0.6	0.8	1.0	2.0	3.0
$\varepsilon^{-0.628t/\tau_n}$	1	0.9	0.78	0.69	0.60	0.53	0.29	0.15
$\text{Cos}(358° t/\tau_n - 5°56')$	1	0.431	−0.738	−0.88	0.184	1.0	1.0	1.0
$100x/X_0$	100	36.2	−57.7	−60.6	11.0	53.0	29.0	15.0

These results are extended in the next table and plotted on Fig. 3.5(a). As regards zeros, $x/x_0 = 0$ when $(\sqrt{(1-\zeta^2)}2\pi(t/\tau_n) + \beta = \pi/2 + n\pi$, where n is any integer, i.e. when $\dfrac{t}{\tau_n} = \dfrac{0.5 + n - \beta/\pi}{2\sqrt{(1-\zeta^2)}} = 0.5025n + 0.267$. Thus, $\dfrac{x}{X_0} = 0$ when $\dfrac{t}{\tau_n} = 0.267, 0.769,$ 1.272, 1.775, 2.277, etc. As regards maxima and minima, $\dfrac{\dot{x}}{X_0} = 0$ when $\dfrac{t}{\tau_n} = 0.5025\,n$, where n is any integer. Thus, maxima and minima occur when $\dfrac{t}{\tau_n} = 0.502, 1.005, 1.507,$ 2.01, etc, with corresponding values of $\dfrac{x}{X} \times 100 = -72.9, 53.2, -38.8, 23.4$, etc. These values are plotted on Fig. 3.5(a) for $\zeta = 0.1$ together with others in the category $0 < \zeta < 1$ in accordance with the following table of calculated values.

	t/τ_n	0	0.25	0.5	0.75	1.0	1.25	1.5	1.75	2.0	2.25	2.5	3.0
$\zeta - 0.2$	$100x/X_0$	100	17.4	−52.4	−11.9	27.3	7.7	−14.1	−4.8	7.3	2.9	−3.7	1.9
$\zeta = 0.6$	$100x/X_0$	100	39.8	−5.6	−7.3	−0.9	0.9	0.35	−0.05		faded out		

(b) $\boldsymbol{\zeta = 1}$, $\dfrac{x}{X} = (1 + \omega_n t)\varepsilon^{-\omega_n t}$—(as in Appendix B12(iii)). This may alternatively be written $\dfrac{x}{X} = (1 + 2\pi t/\tau_n)\varepsilon^{-2\pi t/\tau_n}$ from which we deduce that when $t/\tau_n = 0.5$, then $\dfrac{x}{X_0}$ $= (1+\pi)\varepsilon^{-\pi} = 0.179$ and the following values similarly.

	t/τ_n	0	0.25	0.5	0.75	1.0	1.25
$\zeta = 1.0$	$100x/X_0$	100	53.3	17.9	5.15	1.4	0.35

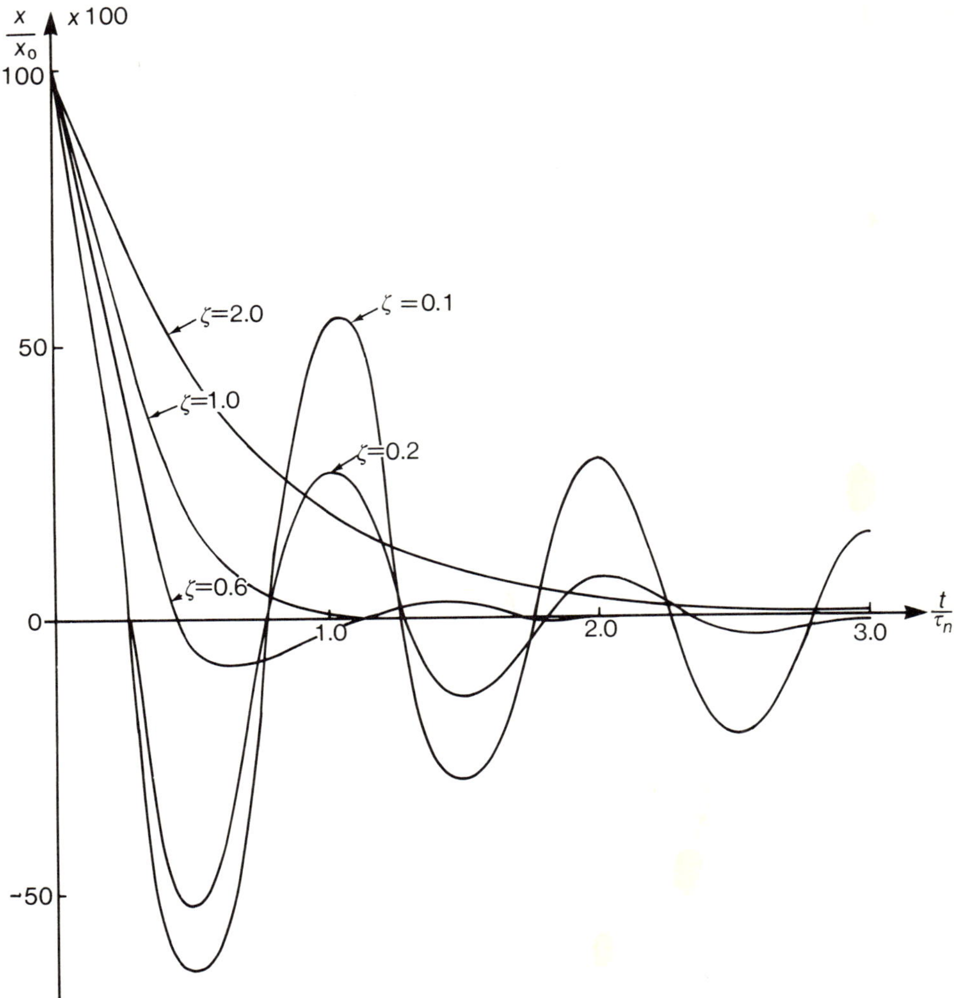

Fig. 3.5(a) Non-dimensional curves of damped natural oscillations

These are plotted on Fig. 3.5(a) where it will be seen that when $\zeta = 1$ the mass just fails to overshoot as it returns to the equilibrium position.
(c) $\boldsymbol{\zeta > 1}$**,** $x = C_1 \varepsilon^{(-\zeta + \sqrt{(\zeta^2 - 1)})\omega_n t} + C_2 \varepsilon^{(-\zeta - \sqrt{(\zeta^2 - 1)})\omega_n t}$. When $t = 0$, $x = X_0 = 100$ divisions and $\dot{x} = 0$, i.e.

$$X_0 = C_1 + C_2 \text{ and } (-\zeta + \sqrt{(\zeta^2 - 1)})C_1 + (-\zeta - \sqrt{(\zeta^2 - 1)})C_2 = 0.$$

Hence,

$$\frac{x}{X_0} = \frac{(\zeta + \sqrt{(\zeta^2 - 1)})}{2\sqrt{(\zeta^2 - 1)}} \varepsilon^{(-\zeta + \sqrt{(\zeta^2 - 1)})\omega_n t} + \frac{(-\zeta + \sqrt{(\zeta^2 - 1)})}{2\sqrt{(\zeta^2 - 1)}} \varepsilon^{(-\zeta - \sqrt{(\zeta^2 - 1)})\omega_n t}.$$

In particular, when $\zeta = 2.0$, $\dfrac{x}{X_0} = 1.08\varepsilon^{-0.268 \times 2\pi t/\tau_n} - 0.0774\,\varepsilon^{-3.73 \times 2\pi t/\tau_n}$ from which we deduce that when $t/\tau_n = 1$, $x/X_0 = 0.20$ or $100x/X_0 = 20$, and the following values similarly

	t/τ_n	0	0.25	0.5	0.75	1.0	1.25	1.5	2.0	2.5	3.0
$\zeta = 2.0$	$100x/X_0$	100	70.8	45	30	20	13.1	8.6	3.7	1.61	0.69

These are plotted on Fig. 3.5(a) showing dead-beat motion.

3.6 LOGARITHMIC DECREMENT

Successive amplitudes of a viscously-damped harmonic motion (see Fig. 3.6(a)) have a logarithmic relationship with one another. Hence, a convenient way of determining the amount of damping present in such a system is to measure the rate of decay of oscillation—best expressed by the term 'logarithmic decrement' defined as the natural logarithm of the ratio of any two successive amplitudes, i.e. $\delta = \log_\varepsilon(x_1/x_2)$ as in Fig. 3.6(c)—a typical value for, say, shock absorbers of automobiles being about 2.

In cases of 'light' damping ($0 < \zeta < 1$ of Section 3.3)

$$x = C\varepsilon^{-\zeta\omega_n t}\cos(\omega_d t + \beta) = X\cos(\omega_d t + \beta),$$

where $\omega_n = \sqrt{(k/m)}$, $\omega_d = \omega_n\sqrt{(1-\zeta^2)}$ and $\zeta = c/c_c = (c/2)\sqrt{(mk)}$. When

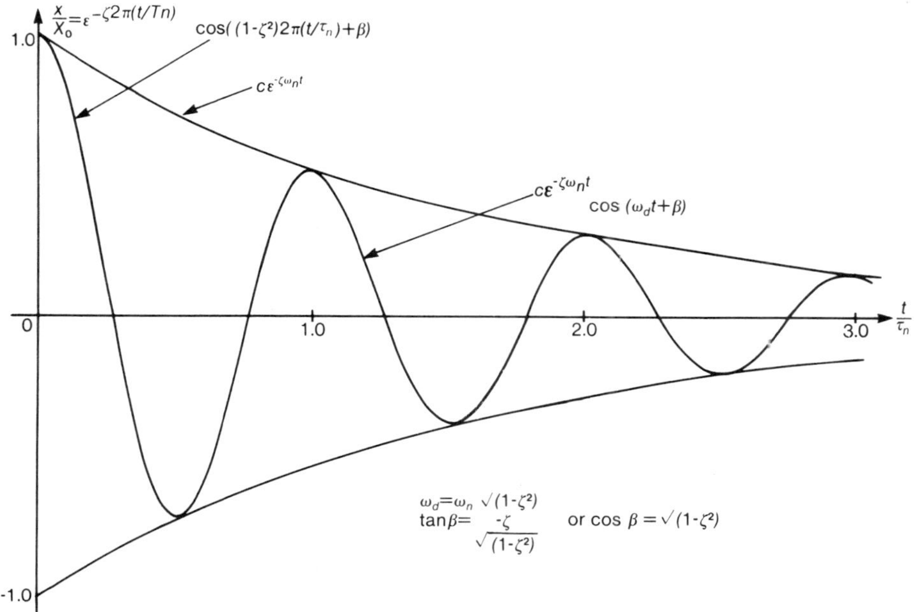

Fig. 3.6(a) Non-dimensional displacement-time graph of damped ($\zeta = 0.1$) natural vibration

$\cos(\omega_d t + \beta) = 1.0$, the curve of x against t is almost tangential to the exponential envelope $C\varepsilon^{-\zeta\omega_n t}$—the tangents not being horizontal, and the points of tangency are slightly to the right of maximum amplitudes. Fig. 3.6(a) shows the time-dependent behaviour of such systems. The decrement factor is x_1/x_2 which, since $x_2 = C\varepsilon^{-\zeta\omega_n t_2}\cos(\omega_d t_1 + 2\pi + \beta) = C\varepsilon^{-\zeta\omega_n t_2}\cos(\omega_d t_1 + \beta)$, can be written $\dfrac{x_1}{x_2} = \varepsilon^{-\zeta\omega_n(t_1 - t_2)} = \varepsilon^{+\zeta\omega_n\tau_d}$. Hence, the logarithmic decrement is $\delta = \log_\varepsilon\left(\dfrac{x_1}{x_2}\right) = \zeta\omega_n\tau_d = \dfrac{2\pi\zeta}{\sqrt{(1-\zeta^2)}}$, which is independent of amplitude.

For small values of ζ, the logarithmic decrement $\delta \doteqdot 2\pi\zeta$, i.e. $\delta \propto \zeta$ (see Fig. 3.6(b)). Also, for small values of ζ,

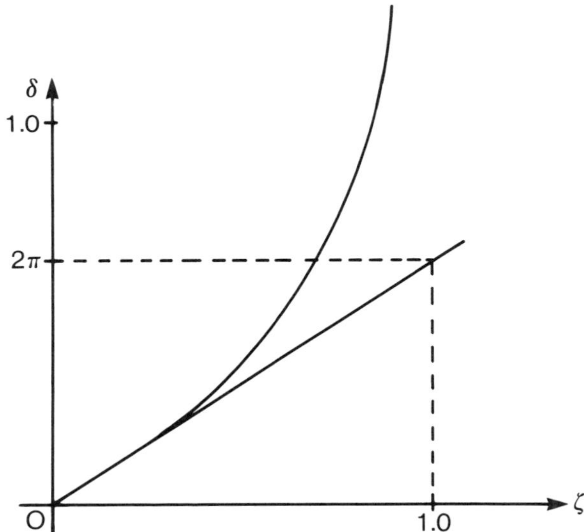

Fig. 3.6(b) Logarithmic decrement against damping factor

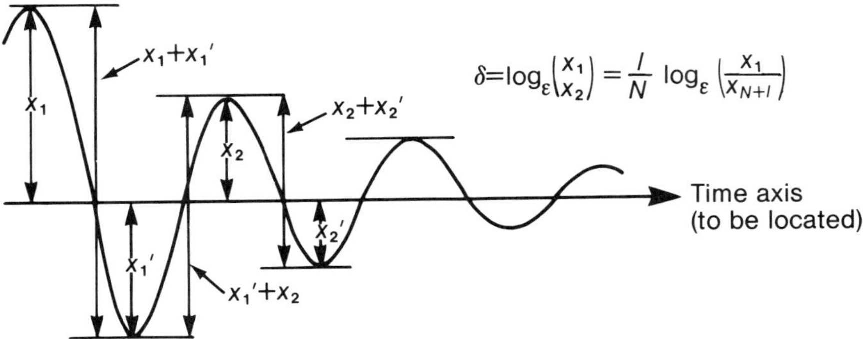

Fig. 3.6(c) Estimation of logarithmic decrement from the 'signature' of a vibration

$$\delta = \log_\varepsilon\left(\frac{x_1}{x_2}\right) = \log_\varepsilon\left(1 + \frac{\Delta x_2}{x_2}\right) \doteqdot \frac{\Delta x_2}{x_2} \doteqdot \frac{\Delta x}{x}$$

(see Fig. 3.6(a)) since $\log_\varepsilon(1 + a) = a - a^2/2 + a^3/3 - a^4/4 + \cdots$ (as in Appendix B5).

This enables a direct approximate value of δ to be made from simple measurements of the damped oscillation curve (which may be regarded as the 'signature' of a vibration) by first drawing the time-line correctly such that in Fig. 3.6(c) $\dfrac{x_1}{x_1^1} = \dfrac{x_1^1}{x_2} = \dfrac{x_2}{x_2^1} = \cdots = \dfrac{x_{N+1}}{x'_{N+1}}$.

Then $\dfrac{x_1 + x_1^1}{x_1^1} = \dfrac{x_1^1 + x_2}{x_2}$ and $\dfrac{x_1^1}{x_2} = \dfrac{x_1 + x_1^1}{x_1^1 + x_2} = \dfrac{x_1}{x_1^1}$. Similarly, $\dfrac{x_1^1}{x_2} = \dfrac{x_1^1 + x_1}{x_2 + x_1^1}$ up to the $(N + 1)^{th}$ cycle, where $\dfrac{x_N^1}{x_{N+1}} = \dfrac{x_N^1 + x_N}{x_{N+1} + x_N^1}$. Hence, since

$$\frac{x_1}{x_1^1} \times \frac{x_1^1}{x_2} \times \frac{x_2}{x_2^1} \times \frac{x_2^1}{x_3} \ldots \times \frac{x_N}{x_N^1} \times \frac{x_N^1}{x_{N+1}} = \left(\frac{x_1}{x_1^1}\right)^{2N} = \left(\frac{x_1}{x_2}\right)^N,$$

the logarithmic decrement and ζ can be estimated over N cycles from $\delta = \log_\varepsilon\left(\dfrac{x_1}{x_2}\right) = \dfrac{1}{N}\log_\varepsilon\left(\dfrac{x_1}{x_{N+1}}\right)$ and $\delta = \zeta\omega_n\tau_d = \dfrac{2\pi\zeta}{\sqrt{(1 - \zeta^2)}}$ (see Fig. 3.6(c)). Alternatively, since $\dfrac{x_1}{x_2} = \left(\dfrac{x_1}{x_1^1}\right)^2, \delta = \log\left(\dfrac{x_1}{x_1^1}\right) = \pi\zeta/\sqrt{(1 - \zeta^2)}$.

In an m, k, c system the variables are ω_n and ζ, and if we want to know them for an instrument, say, we may displace the system from its equilibrium position and allow it to draw its 'signature' whilst comming to rest. Measure τ_d and the relevant values of x to estimate δ and ζ and, hence, ω_n from $2\pi/\tau_d = \omega_d = \omega_n\sqrt{(1 - \zeta^2)}$ after which the motion can be predicted for various values of the dimensionless damping factor ζ. Logarithmic decrement may also be expressed as half the ratio of the energy dissipated to the total vibrational energy of the system for, assumming Hooke's law to be obeyed, the total vibrational energy is proportional to the square of the amplitude, i.e. $E = qx^2$, where q is a constant of the system. To determine the energy dissipated in successive cycles, we consider the vibrational energy a cycle later, i.e. $E - \Delta E = q(x - \Delta x)^2$ in accordance with Fig. 3.6(a). Hence, neglecting Δx^2, we deduce that $E - \Delta E = q(x^2 - 2x\Delta x) = E - 2qx\Delta x$ and $\dfrac{\Delta E}{E} = \dfrac{2qx\Delta x}{qx^2} = \dfrac{2\Delta x}{x}$. Therefore since, for small values of ζ, $\delta \doteqdot \dfrac{\Delta \dot{x}}{x}$ the logarithmic decrement $\delta \doteqdot \dfrac{1}{2}\dfrac{\Delta E}{E}$, which enables a percentage estimation of the energy dissipated in each successive cycle to be made when δ is known.

Another specification of the damping in a freely-vibrating system is the time required for the amplitude to decay to a certain fraction of its initial value (see Example 3.6(ii)).

EXAMPLE 3.6(i)—illustrative of reduction of amplitude by damping
Data for a vibrating system with viscous damping are $m = 4.5$ kg, $k = 5.0$ N/mm, $c = 21.0$ Ns/m. Calculate the damping factor, the logarithmic decrement and the ratio of successive amplitudes. Also, deduce the frequency of damped oscillation.

Undamped natural angular frequency

$$\omega_n = \sqrt{(k/m)} = \sqrt{\left(\frac{5N \times 10^3}{4.5\,\text{kg m}}\right)\left[\frac{\text{kg m}}{\text{Ns}^2}\right]} = \frac{100}{3\text{s}} \text{or } 38.3 \text{ rad/s}.$$

Damping factor $\zeta = \dfrac{c}{2m\omega_n} = \dfrac{21\text{ Ns}}{100\text{ m}} \times \dfrac{3\text{s}}{9\text{ kg}}\left[\dfrac{\text{kg m}}{\text{N s}^2}\right] = 0.07$. Logarithmic decrement $\delta = \dfrac{2\pi\zeta}{\sqrt{(1-\zeta^2)}} = \dfrac{0.14\,\pi}{\sqrt{0.9951}} = 0.442$. Since $\delta = \log_\varepsilon\left(\dfrac{x_1}{x_2}\right) = 0.442$, the amplitude ratio for any two consecutive cycles is $\dfrac{x_1}{x_2} = \varepsilon^\delta = 1.56$, and since $\omega_d = \omega_n\sqrt{(1-\zeta^2)} = \dfrac{100}{3\text{s}}\sqrt{0.995} = 33.2/\text{s}$, then $f_d = \dfrac{\omega_d}{2\pi} = 5.3$ cycles/s.

EXAMPLE 3.6(ii)—illustrative of cyclic time for 'half-life' oscillation
Deduce a formula giving time (t_h) of endurance and the number (N_h) of cycles to 'half-life' of a vibration (i.e. $0 < \zeta < 1$, necessarily).

Referring to Fig. 3.6(c) and deductions $\delta = \dfrac{1}{N}\log_\varepsilon\left(\dfrac{x_1}{x_{N+1}}\right) = \zeta\omega_n\tau_d$ and $t_h = N_h\tau_d$, then $t_h = \dfrac{\log_\varepsilon 2}{\zeta\omega_n}$ showing that only ζ and ω_n are required for estimating the time of endurance to half-life. Also, when $\dfrac{x_1}{x_{N+1}} = \dfrac{1}{2}$, $N_h\tau_d = \dfrac{\log_\varepsilon 2}{\zeta\omega_n}$. Hence, $\dfrac{\log_\varepsilon 2}{\zeta\omega_n\tau_d} = \dfrac{\log_\varepsilon 2}{2\pi}\left(\dfrac{\sqrt{(1-\zeta^2)}}{\zeta}\right) = \dfrac{\log_\varepsilon 2}{2\pi\tan\psi}$ in accordance with Fig. 3.3(c). Thus, approximately, the number of cycles required to elapse for a 50 per cent reduction in amplitude is given by the hyperbolic law $N_h\zeta = \dfrac{\log_\varepsilon 2}{2\pi} = 0.110$. Thus, in Example 3.6(i), 'half-life' occurs after $N_h = \dfrac{0.110}{0.07} = 1.57$ cycles.

3.7 LABORATORY WORK ON DAMPED NATURAL OSCILLATIONS

Experiment 1. Object: (i) to find the virtual moment of inertia (I) of a cylinder oscillating (a) in air and (b) in oil, (ii) to estimate the damping factor (ζ) and the natural frequency (ω_n).

Apparatus and method:- allow the cylinder to oscillate on the end of a wire in torsion (see Fig. 3.7(a)). Then, torque $T = I\ddot{\theta} = -\dfrac{G\theta J}{L}$ (see Fig. A6(a))

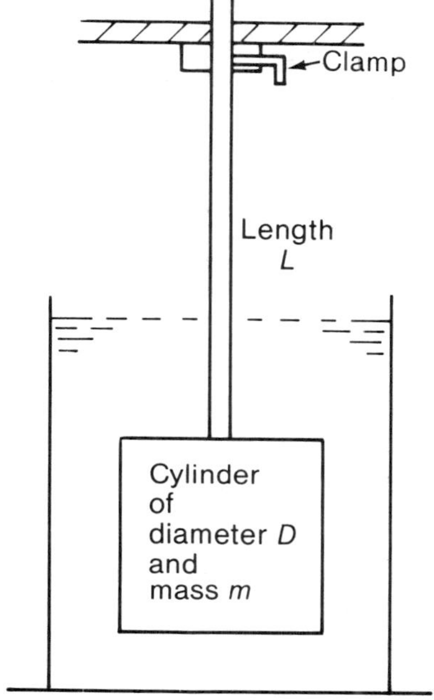

Fig. 3.7(a) Cylinder in torsional oscillation

and $\tau = 2\pi\sqrt{\left(\dfrac{IL}{JG}\right)}$. Calculate $I = mD^2/8$ and estimate G of the wire from $G = \dfrac{IL}{J}\left(\dfrac{2\pi}{\tau}\right)^2$ having found L/τ^2 from a graph of τ^2 plotted against various lengths L.

Hence (i) (a) the virtual I of the cylinder in air will be $mD^2/4$ unless found from $I_{\text{air}} = \dfrac{JG}{4\pi^2}\left(\dfrac{\tau^2}{L}\right)$ after G has been found by a method other than the above, and $\omega_n = \sqrt{\left(\dfrac{GJ}{IL}\right)}$; (b) the virtual I of the cylinder in oil can be estimated from $I_{\text{oil}} = \dfrac{JG}{4\pi^2}\left(\dfrac{\tau^2}{L}\right)$ after τ^2/L has been found from a plot of τ^2 against L with the cylinder oscillating in oil.

(ii) In air and in oil measure with, say, the aid of an optical lever and a scale, the maximum positive amplitudes and record the times

Swings	1	2	3	... N	$N+1$
max. scale divisions	x_1	x_2	x_3	... x_N	x_{N+1}
time	t_1	t_2	t_3	... t_N	t_{N+1}

Plot the decrement curve, x against t, and x against N. Estimate τ_d by noting the time for, say, 10 or 20 cycles. Also, since logarithmic decrement

$$\delta = \log_\varepsilon\left(\frac{x_1}{x_2}\right) = \frac{2\pi\zeta}{\sqrt{(1-\zeta^2)}} = \zeta\omega_n\tau_d = \frac{1}{N}\log_\varepsilon\left(\frac{x_1}{x_{N+1}}\right)$$

(reckoned over 10 cycles) then $\delta = \frac{1}{10}\log_\varepsilon\left(\frac{x_1}{x_{11}}\right) = \frac{2\pi\zeta}{\sqrt{(1-\zeta^2)}}$ and, hence, ζ and $\omega_n = \frac{2\pi}{\tau_d\sqrt{(1-\zeta^2)}}$. Alternatively, since $x_{N+1} = x_1\varepsilon^{-\zeta\omega_n N\tau_d}$ $\log_\varepsilon x_5 - \log_\varepsilon x_{15} = \zeta\omega_n 10\tau_d$ and, hence, ζ since $\omega_n = \sqrt{\left(\frac{GJ}{IL}\right)}$.

Experiment 2. Object: To find the apparent moment of inertia of a four-blade propeller of an outboard motor (a) in air (b) submerged in water—i.e. of the effect of the entrained water and damping on the virtual moment of inertia.

As in Experiment 1 the propeller is allowed to oscillate on the end of a wire in torsion. But first the modulus of rigidity of the wire can be estimated from $G = 4\pi^2\frac{I}{J}\left(\frac{L}{\tau^2}\right)$, where $I = \frac{mD^2}{8}$ is the moment of inertia of say, a lead cylinder of mass m and diameter D.

Then, with the propeller oscillating on the end of the wire in air the moment of inertia of the propeller can be estimated from $I_{air} = \frac{JG}{4\pi^2}\left(\frac{\tau^2}{L}\right)$ and $\omega_n = \sqrt{\left(\frac{GJ}{LI_{air}}\right)}$. Similarly, with the propeller oscillating in water on the end of the same wire, measure the maximum positive amplitudes—say by means of an optical lever and a scale, $I_{water} = \frac{JG}{4\pi^2}\left(\frac{\tau_d^2}{L}\right)$ and from the decrement curve of x against t obeying the law $x = C\varepsilon^{-\zeta\omega_n t}$ or $\log_\varepsilon x = \log_\varepsilon C - \zeta\omega_n t$ we may deduce that $\zeta = \frac{\log_\varepsilon x_1 - \log_\varepsilon x_{21}}{\omega_n(t_{21}-t_1)}$ or $\zeta = \frac{\log_\varepsilon(x_1/x_{21})}{20\omega_n\tau_d}$ since the time taken for 20 cycles, namely, $(t_{21}-t_1) = 20\tau_d$

and, in general, over N cycles, $\zeta = \frac{\log_\varepsilon\left(\frac{x_1}{x_{N+1}}\right)}{\omega_n \times N\tau_d}$.

EXAMPLE 3.7(i)—illustrative of experimental determination of the apparent moment of inertia of a propeller in air and in water

If in Experiment 2 the diameter of the wire is $d = 2.5$ mm (i.e. $J = \pi d^4/32 = 3.834\,\text{mm}^4$)

and the cylinder has a diameter of 80 mm and a mass of 7 kg (i.e. $I = \frac{mD^2}{8} =$ 5600 kg mm^2) and the periodic time for torsional oscillation of the cylinder on the end of the wire of length 600 mm is 0.84 s (i.e. $L/\tau^2 = 841$ mm/s^2) then the modulus of rigidity of the wire is

$$G = \frac{4\pi^2 I}{J}\left(\frac{L}{\tau^2}\right) = \frac{4 \times 9.87}{3.843} \times 5600 \times 841 \frac{\text{kg}}{\text{mm s}^2}\left[\frac{\text{N s}^2}{\text{kg m}}\right] = 48.5\ \text{GN/m}^2$$

and $JG = 0.186$ Nm^2.

Also, if when the propeller of mass 1.5 kg oscillates in air on the end of the same kind of wire of length 967 mm, the periodic time is found to be 0.815 s, then the moment of inertia of the propeller in air is

$$I_a = \frac{JG}{4\pi^2}\left(\frac{\tau^2}{L}\right) = \frac{0.186\ \text{Nm}^2}{4 \times 9.87}\left(\frac{0.815^2\ \text{s}^2}{0.967\ \text{m}}\right)\left[\frac{\text{kg m}}{\text{N s}^2}\right]$$
$$= 3.24\ \text{gm}^2 \text{ or } 3240\ \text{kg mm}^2$$

and the angular frequency of natural undamped or free oscillation is

$$\omega_n = \sqrt{\left(\frac{JG}{I_a L}\right)} = \sqrt{\left(\frac{0.186\ \text{Nm}^2}{0.00324\ \text{kgm}^2 \times 0.967\ \text{m}}\right)} = \sqrt{(59.3/\text{s}^2)} = 7.7/\text{s}$$

Similarly if, when the propeller oscillates submerged in water on the end of that wire of length 1272 mm, the periodic time is 1.04s, then the moment of inertia of the propeller in water is $I_w = \frac{JG}{4\pi^2}\left(\frac{\tau^2}{L}\right) = 4.0\ \text{gm}^2$, and the increase in virtual moment of inertia is $\frac{0.76}{3.24}$ or 23.5%.

From decrement observations on number of swings and maximum positive amplitude when oscillating in water it was found that over 10 cycles the ratio of these amplitudes $\left(\frac{x_1}{x_{11}}, \frac{x_5}{x_{15}}, \frac{x_7}{x_{17}}\right)$ was 2.4. Hence, since

$$\delta = \frac{1}{N}\log_\varepsilon\left(\frac{x_1}{x_{N+1}}\right) = \zeta\omega_n\tau_d, \text{ then } \zeta = \frac{\log_\varepsilon 2.4}{10 \times 1.04 \times 6.97} = 0.012$$

Alternatively, $\delta = \frac{1}{N}\log_\varepsilon\left(\frac{x_1}{x_{N+1}}\right) = \frac{2\pi\zeta}{\sqrt{(1-\zeta^2)}} = \frac{1}{10}\log_\varepsilon(2.4) = 0.08755.$ Hence, $\zeta = 0.014$ showing a discrepancy of 0.002.

3.8 NATURAL VIBRATION WITH COULOMB OR HYSTERETIC DAMPING

Because Coulomb investigated dry friction between moving bodies two centuries ago, his name is associated with dry friction damping which is independent of velocity, i.e. frictional force $F = \mu R$ is independent of displacement x and its derivatives $\dot{x}$ and $\ddot{x}$.

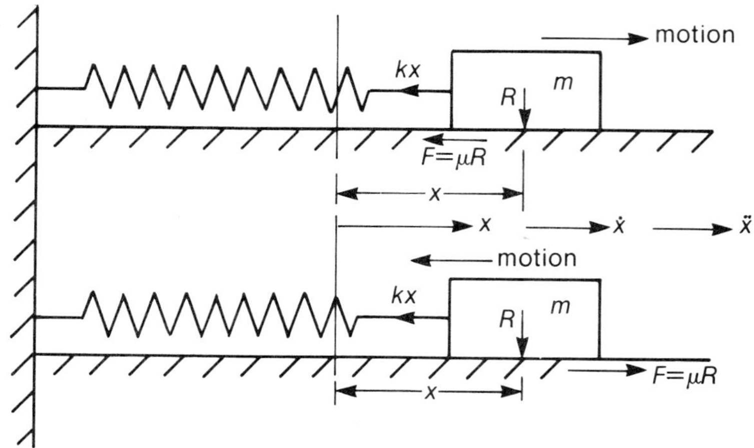

Fig. 3.8(a) Oscillation with constant damping

Referring to Fig. 3.8(a), it will be seen that we cannot take account of the damping force for the complete cycle in one equation of motion because, since friction opposes motion, the sign of the friction force changes when the direction of motion changes. Hence, two equations are necessary, namely:

(i) when motion is from left to right, $m\ddot{x} = -kx - F$

(ii) when motion is from right to left, $m\ddot{x} = -kx + F$

Thus, in the latter case, $\ddot{x} + \frac{k}{m}x = [D^2 + \omega_n^2]x = \frac{F}{m}$, the solution of which is composed of the complementary function $x = A \sin \omega_n t + B \cos \omega_n t = C \cos(\omega_n t + \beta)$ and the particular integral $\frac{F}{k}\left(\text{since} \frac{F}{k} \times \frac{k}{m} = \frac{F}{m}\right)$ making the complete solution $x = A \sin \omega_n t + B \cos \omega_n t + F/k$ (see Appendix B12 and 13).

If the motion is started with zero velocity and the initial displacement x_0 then $(\dot{x})_{t=0} = \omega_n A = 0$, and $(x)_{t=0} = x_0 = B + F/k$. Hence, $x = (x_0 - F/k) \cos \omega_n t + F/k$ which applies only when the mass is moving to the left.

We may note that a plot of this motion could be generated using a micro-computer. The last equation would translate into a basic language statement X = (XØ − F/k)∗cos(W∗T) + (F/k)∗SGN(V) in which V must first be computed from V = −(XØ − F/k)∗W∗SIN(WT). Since, however, we are only interested in SGN(V), it would be sufficient to compute SGN(SIN(W∗T)). In this computer language, W represents omega (ω), T time (t) and K stiffness (k).

The amplitude of the motion at the extreme left occurs when $\omega_0 t = \pi$, i.e. $x = (x_0 - F/k)(-1.0) + F/k = -(x_0 - 2F/k)$ showing that the amplitude at the end of the first half-cycle has been diminished by $2F/k$. Similarly,

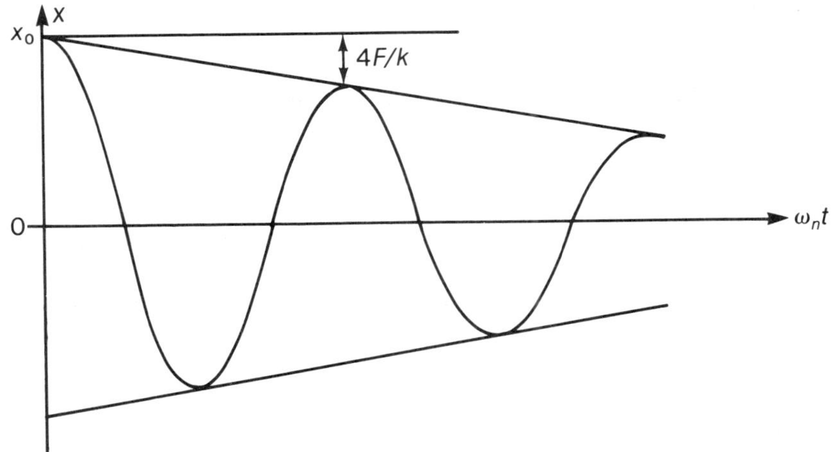

Fig. 3.8(b) Effect of constant damping

motion from left to right causes a further diminution of amplitude of $2F/k$. Hence, the decay per cycle is $4F/k$, and the envelope of the damped curve is a straight line as shown in Fig. 3.8(b). Motion will cease when the maximum amplitude is less than Δ, i.e. the position at which the spring for $kx = k\Delta$ is unsufficient to overcome the static friction force $F = \mu R$. The periodic time, $\tau = 2\pi/\omega_n$, is the same as that of an undamped system, i.e. constant damping does not change the period of vibration.

EXAMPLE 3.8(i)—illustrative of amplitude reduction by constant or hysteretic damping

Referring to Fig. 3.8(a), if $m = 4.5$ kg, $\mu = 0.025$ and $k = 0.90$ N/m, find the frequency of oscillations and the number of cycles corresponding to 50 per cent reduction in amplitude when the initial amplitude is 49 mm.

Natural angular frequency $\omega_n = \sqrt{(k/m)}$. Hence, in this case,

$$\omega_n = \sqrt{\left(\frac{0.90\,\text{N}}{4.5\,\text{kg mm}}\left[\frac{\text{kg}\,10^3\text{mm}}{\text{N}\,\text{s}^2}\right]\right)}$$

$$= 14.12/\text{s}.$$

Alternatively, $f_n = 14.12\dfrac{\text{rad}}{\text{s}}\left[\dfrac{\text{cycle}}{2\pi\,\text{rad}}\right] = 2.25$ cycle/s. Reduction in amplitude per cycle is

$$\frac{4F}{k} = 4\frac{\mu mg}{k} = \frac{4 \times 0.025 \times 4.5}{0.90\,\text{N/mm}}\text{kg} \times 9.81\frac{\text{m}}{\text{s}^2} = 4.9\,\text{mm}$$

Hence, the initial amplitude is reduced by 24.5 mm in 5 cycles.

Hysteretic or solid damping caused, say, by internal friction in material of structural members experiencing vibratory motion is due to the cyclic manner in which strain changes with stress—the enclosed area of the hysteresis loop being proportional to the energy dissipated during the cycle and depends on the material. This has to be dealt with by a method other than a differential equation of motion since such an equation will not indicate the direction (see Fig. 3.8(a)) of the damping forces throughout a cycle.

Assuming the damping force (f_d) is proportional to displacement (x) and independent of frequency, i.e. $f_d = \gamma kx$, where k represents stiffness and γ is the (non-dimensional) constant of proportionality, the energy dissipated per cycle is approximately $\pi\gamma kX^2$ for harmonic motion of amplitude X (see Example 3.1(i)).

Another mathematical ploy (difficult to substantiate physically) is to assume internal damping force (f_d) is proportional to velocity ($\dot{x}$) and inversely proportional to angular frequency (ω)—i.e. $f_d = h\dot{x}/\omega$, where h is the hysteretic damping constant as shown diagrammatically in Fig. 3.8(c). This leads to the differential equation

$$f_d = kx + h\dot{x}/\omega = -m\ddot{x}$$

or

$$\ddot{x} + \frac{h}{\omega m}\dot{x} + \frac{k}{m}x = 0,$$

i.e. in which h/ω replaces c of the previous equations relating to viscous damping.

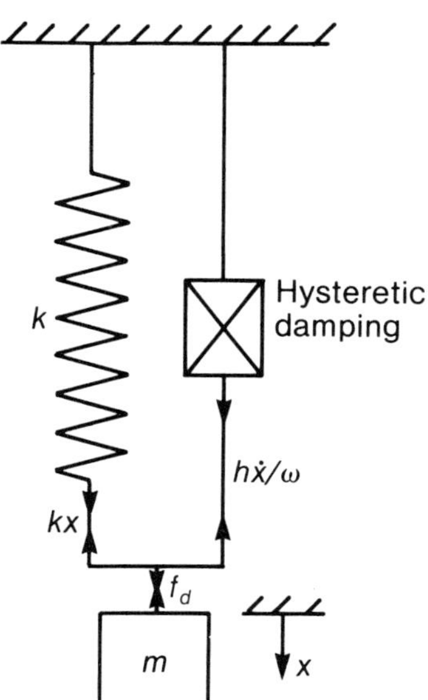

Fig. 3.8(c) Hysteretic damping

EXAMPLE 3.8(ii)—illustrative of equivalent viscous damping
In cases where the actual damping force is assumed proportional to the nth power of the relative velocity.

Actual damping in practical systems is usually a combination of different types, but its overall effect can be represented and included in analysis by substituting an equivalent viscous damper (i.e. whose resisting force is proportional to relative velocity) in the mathematical or idealized model with equivalent energy dissipation. Thus, if the relative displacement between the sliding parts of a viscous damper is $x = X\cos\omega t$, their relative velocity will be $\dot{x} = -\omega X \sin\omega t$ and the force exerted against the motion is, by definition, $c\dot{x}$, where c is the coefficient of viscous damping. The work done by this force during an increment of displacement δx (opposite in direction to the force) is $\delta W = c\dot{x}\,\delta x = cX^2\omega^2 \sin\omega t\,\delta t$.

If, however, the force exerted is assumed to be $c_n(\dot{x})$, where c_n is the coefficient of actual damping, then the energy dissipated is $\delta W_n = c_n(\dot{x})^n\delta x = c_n(-X\omega\sin\omega t)^{n+1}\delta t$, and an equivalent viscous damping coefficient c or c_e (causing the same amount of energy dissipation as the actual damping) can be found by equating δW to δW_n and integrating over a quarter of a cycle of motion. Thus, since $2\sin^2\omega t = 1 - \cos 2\omega t$ (see Appendix B5).

$$c_e X^2\omega^2 \int_0^{\pi/2\omega} \left(\frac{1-\cos 2\omega t}{2\omega}\right) \mathrm{d}(\omega t) = \frac{c_n}{\omega}(-X\omega)^{n+1} \int_0^{\pi/2\omega} (\sin\omega t)^{n+1}\,\mathrm{d}\omega t$$

i.e. $\frac{\pi}{4}c_e X^2\omega =$ a quantity depending on n. In particular, (a) if $n = 1$, $c_e = c_1 = c$ as expected for viscous damping; (b) if $n = 2$,

$$\begin{aligned}\frac{\pi}{4}c_e X^2\omega &= c_2 X^3\omega^2 \int_0^{\pi/2\omega} -\sin^3\omega t\,\mathrm{d}\omega t \\ &= c_2 X^3\omega^2 \int_0^{\pi/2\omega} (\cos^2\omega t - 1)\,\mathrm{d}\cos\omega t \\ &= \tfrac{2}{3}c_2 X^3\omega^2 \\ \text{i.e. } c_e &= \frac{8}{3\pi}c_2 X\omega\end{aligned}$$

(c) if $n = 0$ (i.e. coulomb or dry friction $f = \mu R = c_0$) and $c_e = \frac{4}{\pi}\frac{c_0}{X\omega}$.

For suggested further reading and study, see Appendix D, (4) and (5).

EXERCISES ON CHAPTER 3

1. The plunger of a dashpot has a mass of 0.25 kg and descends under its own weight in its cylinder full of oil at the rate of 30 mm/s. Calculate the damping factor ζ when the dashpot is used to damp the vibration of an additional mass of 1 kg attached to the end of a spring of stiffness 16 N/mm.
2. Assuming the angle of twist $\theta = \varepsilon^{\omega t}$ deduce an expression for the damping factor of a torsional system having a moment of inertia I, shaft stiffness in torsion k and damping coefficient c.

3. A spring-mass system with viscous damping is displaced a distance x_0 and then released. Determine the equations of motion when the damping factor ζ is 0.2, 1.0 and 2.0. Compare the three cases by plotting non-dimension curves with x/x_0 as ordinate and $\omega_n t$ as abscissa.
4. An instrument with spring control and viscous damping has a uniformly divided scale with a central zero. The pointer is held steady at a reading of +100 and then suddenly released. Plot curves of: (a) the readings of the instrument on a time base from $t = 0$ to 3 s for values of $\zeta = 0$, 0.1, 0.4, 1.0, 5.0 assuming that the natural undamped period is 1 s. (b) What is the fractional loss of amplitude per cycle when $\zeta = 0.05$. (c) Find an expression for semi-life of a damped oscillation and its value when $\zeta = 0.05$.
5. Define the term "logarithmic decrement" and deduce a formula for it in terms of the damping factor.
6. A gun barrel of mass 545 kg has a recoil spring of stiffness 292 kN/m. If the barrel recoils 1.22 m on firing, calculate (a) the initial recoil velocity of the barrel, (b) the critical damping coefficient of a dashpot which is engaged at the end of the recoil stroke, and (c) the time required for the barrel to return to a position 50 mm from its initial position.
7. A vibrating system consists of a mass of 2 kg and a spring of stiffness 1.8 N/mm viscously damped such that two consecutive amplitudes are 10 mm and 9.8 mm. Determine the natural cyclic frequency f_n, the logarithmic decrement δ, the damping factor ζ and the damping coefficient c.
8. A body vibrating in a viscous medium has a period of 0.20 s and an initial amplitude of 25 mm. Calculate the logarithmic decrement, the damping factor and the undamped natural angular frequency if the amplitude after 10 cycles is 0.5 mm.
9. A vibrating system has the following constants: $m = 17.5$ kg, $k = 7$ N/mm and $c =$ 70 Ns/m. Estimate the damping factor ζ, the natural cyclic frequency of damped oscillation f_d, the logarithmic decrement δ and the ratio (x_1/x_2) of any two consecutive amplitudes.
10. Find the semi- or half-life of a natural vibration in an m, k, c system when $\tau_d = 1$ s and $\zeta = 0.01$.
11. A four-blade propeller of mass 2 kg and a radius of gyration 100 mm is to be suspended on a single steel wire of 1.5 mm diameter and $G = 80$ GN/m^2 so that the propeller oscillates in its plane of rotation with a period of 2 s in air. Calculate the length of the wire required. If then the propeller is arranged to oscillate submerged in a bath of oil and loses 80 per cent of its amplitude in each successive cycle, estimate the damping factor, the periodic time of damped oscillations and the virtual moment of inertia of the propeller.

4

Forced vibrations with one degree of freedom

Principles and analyses developed in this chapter help to solve many practical problems of forced vibrations, e.g. isolation of engines, compressors and machinery from foundations, instrument panels and frameworks generally.

4.1 UNDAMPED FORCED VIBRATIONS WITH ONE DEGREE OF FREEDOM

Not only forced vibrations, but natural vibrations also are produced by a disturbing force which may range from a simple impulse to a harmonic force or complex function of time. However, after a time the natural (transient) vibration will be damped out by various resistances. Continuing or steady-state oscillation, having the frequency of the impressed periodic force, will remain although, at the beginning of the motion, the transient vibration may have a practical importance, e.g. concerning critical speeds and consequent high stresses.

Undamped forced oscillations are fairly easy to deal with mathematically but, as we shall now appreciate, the vector or phase diagram has its greatest value in making the solution of problems of damped forced oscillations relatively easy. Although, in practice, there is always some damping, hypothetical cases of forced vibrations assuming zero damping are of interest because they yield pointers of practical value. The simplest cases are constrained to a single degree of freedom as shown diagrammatically in Fig. 4.1(a), the applied force, f, being harmonic of frequency ω on the mass, m for which the equation of motion is $m\ddot{x} = mg + f - k(x + d)$ or $\ddot{x} + k/m\ x = f/m = [D^2 + \omega_n^2]x$, where $\omega_n = \sqrt{(k/m)}$, as before.

The complementary function (see Appendix B12), i.e. the transient part, in the solution of this differential equation is $C \cos(\omega_n t + \beta)$; and the parti-

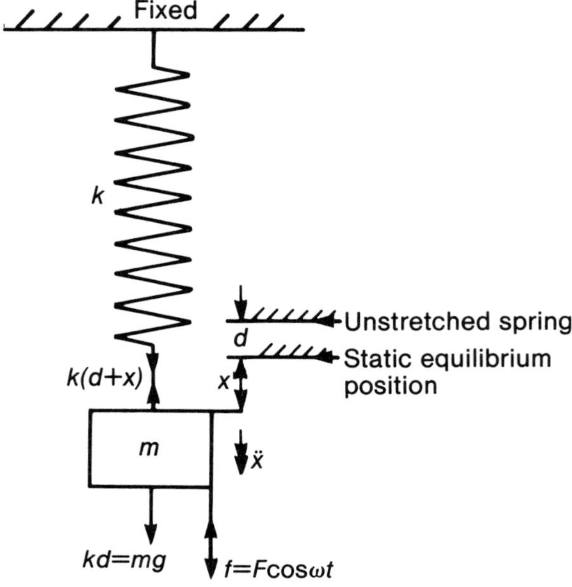

Fig. 4.1(a) Undamped forced oscillations

cular integral, i.e. the steady-state part, is $\dfrac{F\cos\omega t}{m(\mathrm{D}^2+\omega_n^2)} = \dfrac{F\cos\omega t}{m(\omega_n^2-\omega)}$. Thus, the general solution is $x = C\cos(\omega_n + \beta) + \dfrac{F\cos\omega t}{m(\omega_n^2-\omega)}$, provided $\omega_n \neq \omega$, for if $\omega = \omega_n$ (i.e. the frequency of the disturbing force is equal to the frequency of the natural vibration of the system) resonance occurs (as in Example 4.1(ii) and Appendix B13).

The transient term $C\cos(\omega_n t + \beta)$ represents free or natural vibration—it being a motion which could occur without external force (see Section 2.2), and the steady-state term, $\dfrac{F\cos\omega t}{m(\omega_n - \omega)} = X\cos\omega t$, represents forced vibration.

This is a simple harmonic motion having the same periodic time, $\tau = 2\pi/\omega$, as the disturbing or exciting force f of amplitude X which depends on the angular frequency, ω, of the harmonic force f. Hence, $\dfrac{X}{F/k} = \dfrac{1}{1-\left(\dfrac{\omega}{\omega_n}\right)^2} = \dfrac{kX}{F}$ which is the fraction of the maximum disturbing force, F, transmitted to supports, or the 'amplitude ratio' as shown in Fig. 4.1(b). The variation of amplitude (or amplitude ratio) with frequency is called 'the response' of the system, e.g.

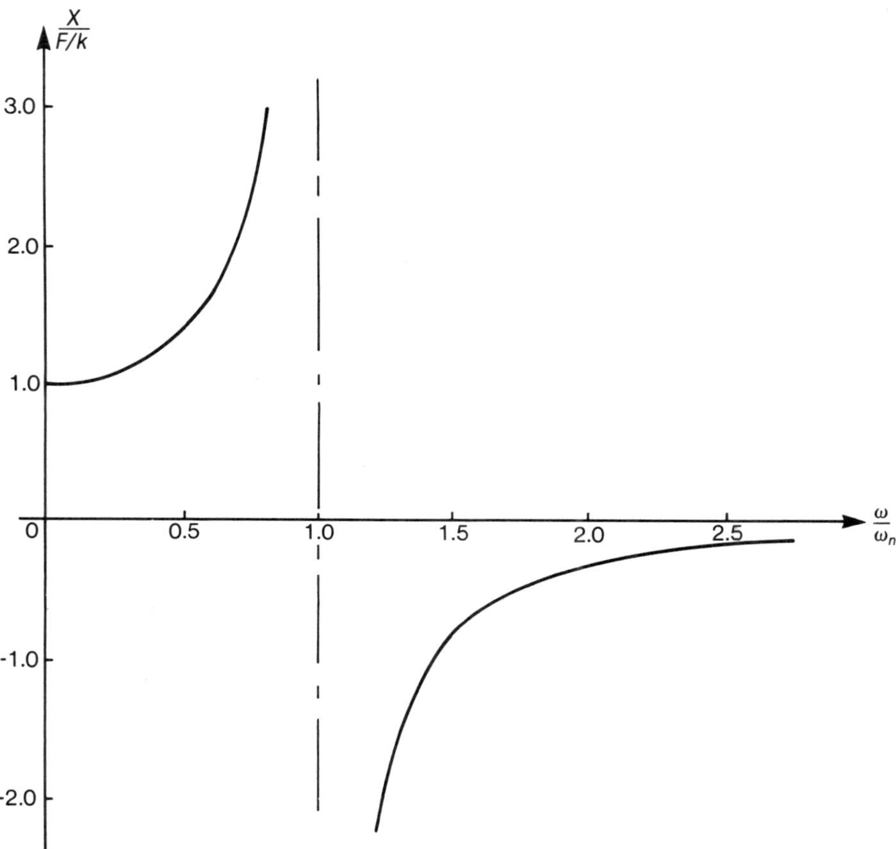

Fig. 4.1(b) Amplitude ratio against frequency ratio

(i) when $\omega < \omega_n$, the amplitude ratio is positive—i.e. the motion is in phase with the applied force;
(ii) when $\omega = \omega_n$, resonance occurs—i.e. the amplitude ratio approaches infinity;
(iii) when $\omega > \omega_n$, the amplitude ratio is negative—i.e. the motion is out of phase with the applied force, f; and
(iv) when $\omega \gg \omega_n$, the amplitude ratio approaches zero—i.e. there is little or no motion of the mass, m.

Alternatively, instead of using the circular function $F \cos \omega t$ for the harmonic force, we may use the complex exponential $F\varepsilon^{j\omega t}$ of which $F \cos \omega t$ is the real part and $F \sin \omega t$ the imaginary part (see Appendix B9). When applied at some point in a dynamic system, this periodic force necessarily induces the system to take up a steady motion with the same frequency, ω, such that (for linear equations of motion) the point of application of the force has the displacement $x = X\varepsilon^{j\omega t} = \alpha F\varepsilon^{j\omega t}$, where $\alpha = X/F$ is termed 'the direct receptance at x' which provides information about the response of a system to a harmonic force. Thus, referring to Fig. 4.1(a), $m\ddot{x} + kx = F\varepsilon^{j\omega t}$ for which a solution (by trial) is $x = X\varepsilon^{j\omega t}$ yielding $X = \dfrac{F}{k - m\omega^2}$ with a direct receptance at x of $\alpha = \dfrac{X}{F} = \dfrac{1}{k - m\omega^2} = \dfrac{1/m}{\omega_n^2 - \omega^2}$. Resonance occurs when α approaches infinity, i.e. when $k - m\omega^2 = 0$ or when $\omega = \sqrt{(k/m)} = \omega_n$.

In the case where the harmonic force is an inertia force caused, say, by a rotating imbalanced mass (i.e. $F(t) = m_0 r\omega^2 \sin \omega t$) as shown diagrammatically in Fig. 4.1(c), the elastic system is represented by supporting springs whose total rating is k, and the whole system is constrained to motion in the vertical direction.

Sideways motion would add a second degree of freedom. Referring to Fig. 4.1(c), and assuming damping to be negligible, the equation of motion of the block of total mass m (including m_0) is $m\ddot{x} = m_0 r\omega^2 \sin \omega t - W - k(x - d)$ in which $W = mg = kd$ and k is the total rating of the supporting springs. Hence, $m\ddot{x} + kx = m_0 r\omega^2 \sin \omega t$ from which we deduce that the steady-state displacement from the static-equilibrium position at any time, t, is $x = \dfrac{m_0 r\omega^2}{m(\omega_n^2 - \omega^2)} \sin \omega t = X \sin \omega t$, where X is the amplitude or maximum displacement. Thus, $\dfrac{mX}{m_0 r} = \dfrac{(\omega/\omega_n)^2}{1 - (\omega/\omega_n)^2}$—shown graphically by the non-dimensional plot of Fig. 4.1(d), from which we see that (i) if $\omega \ll \omega_n$, the amplitude $X \to 0$, (ii) if $\omega = \omega_n$, $X \to \infty$, (iii) if $\omega \gg \omega_n$, the amplitude $X \to -\left(\dfrac{m_0}{m}\right)r$,

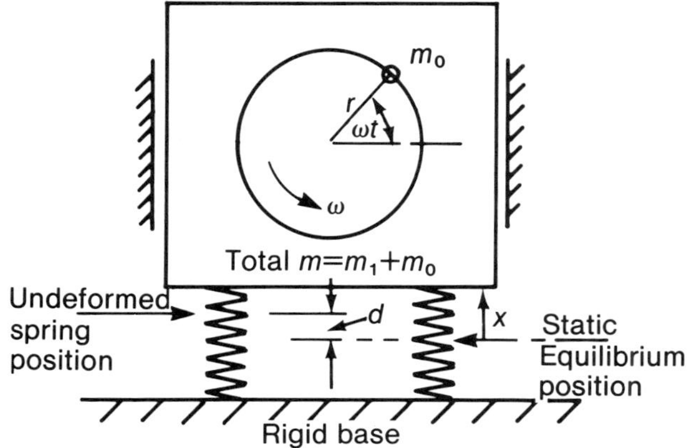

Fig. 4.1(c) Vibration caused by unbalanced rotating mass

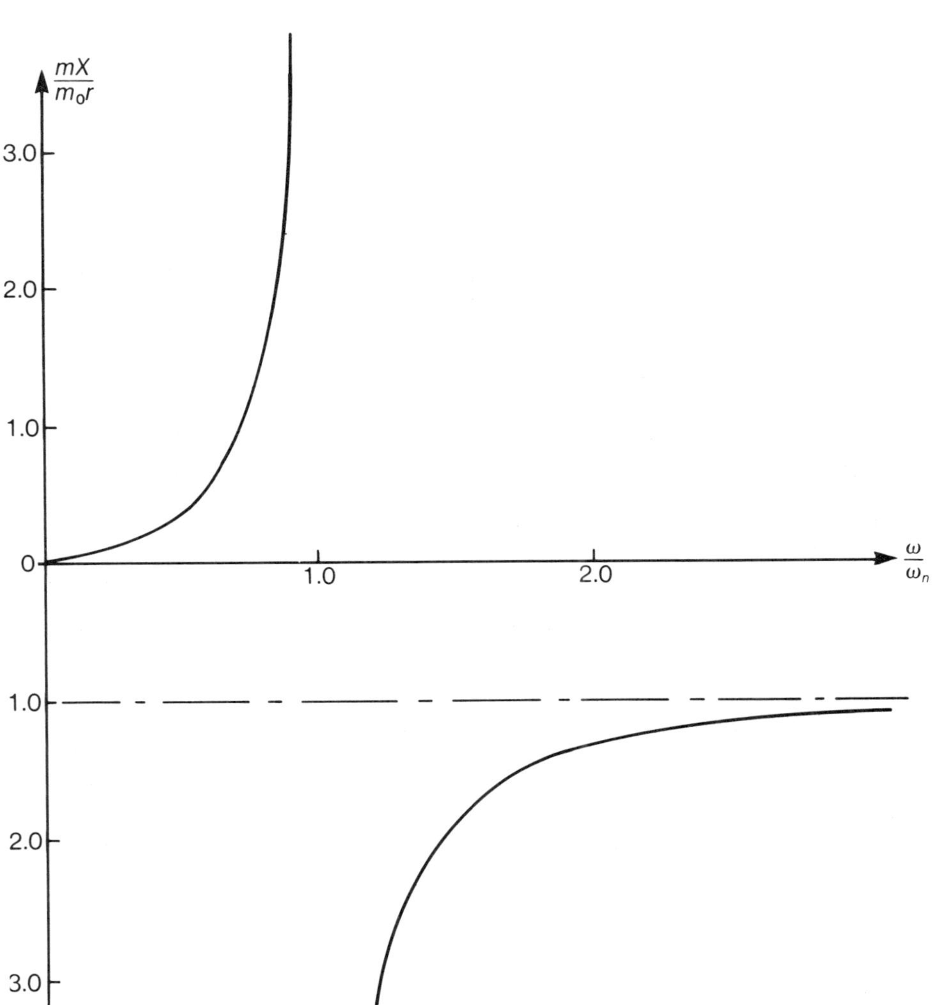

Fig. 4.1(d) Non-dimensional plot of amplitude against frequency ratio

i.e. if the out-of-balance mass, m_o, is a small fraction of the total mass, m, the amplitude, X, of the resulting vibration is proportionally small. Hence, above the speed of resonance, one way of reducing the amplitude, X, of vibrations caused by out-of-balance mass in some mechines is to mount them on relatively heavy frames and so reduce the ratio m_o/m.

EXAMPLE 4.1(i)—illustrative of resonant frequency of a beam
Determine the critical frequency (i.e. when resonance occurs) of a periodic harmonic force applied at mid-span to horizontal beam, neglecting damping.

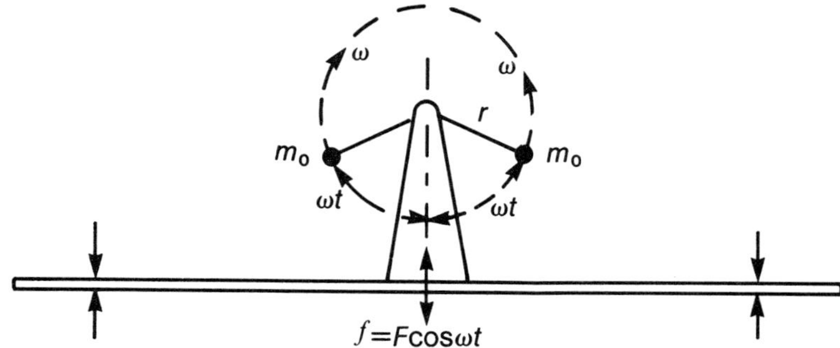

Fig. 4.1(e) Counter-rotating shaker

The harmonic force is applied to the beam of Fig. 4.1(e) by a counter-rotating eccentric-mass exciter (or shaker) such that $f = F \cos \omega t = 2m_o r\omega^2 \cos \omega t$ and the horizontal components balancing. The elastic resistance to vibration is provided by the beam of mid-span stiffness $k = W/d$, where d is the static deflection of the beam under the concentrated load equal to the total weight of the shaker, $W = mg$ (where m includes the eccentric masses, $2m_o$). Hence, the frequency of natural vibration at mid-span is $\omega_n = \sqrt{\left(\frac{k}{m}\right)} = \sqrt{(g/d)}$, and the critical angular frequency of the applied harmonic force is $\omega = \omega_n = \sqrt{(g/d)}$ when resonance occurs in this undamped system.

EXAMPLE 4.1(ii)—illustrative of a build-up to resonance

Consider the case of a simple pendulum whose point of support moves horizontally with simple harmonic motion. Comment on the cases when its frequency is $\langle = \rangle$ that of the natural frequency or free frequency of the simple pendulum.

Referring to Fig. 4.1(f), the point P and mass m at time t are at perpendicular distances $\varepsilon = a \cos \omega t$ and $x = \varepsilon + l\theta$, respectively, from the stationary reference line oo′ for small

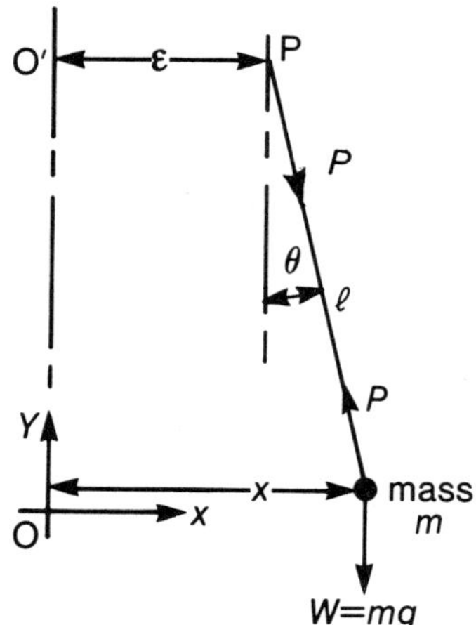

Fig. 4.1(f) Simple pendulum with moving point of support

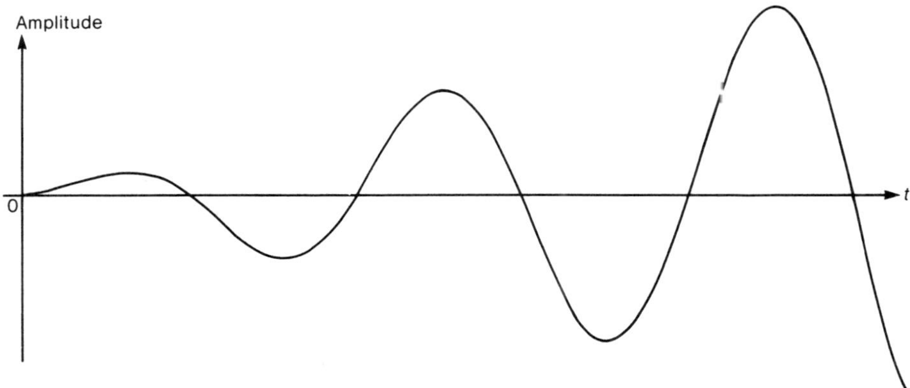

Fig. 4.1(g) Resonance build-up

angles of θ. Hence, since $m\ddot{x} = -p\sin\theta \doteqdot -mg\sin\theta$ and $\sin\theta = (x-\varepsilon)/l$, then $[D^2+\omega_n^2]x = \omega_n^2 a\cos\omega t$, where $\omega_n^2 = g/l$.

The general solution of this differential equation (see Appendix B13) is $x = A\cos\omega_n t + B\sin\omega_n t + \dfrac{\omega_n^2 a}{\omega_n^2-\omega^2}\cos\omega t$ provided that $\omega_n \neq \omega$ (see (iii) following).

The particular integral term, $\dfrac{\omega_n^2 a}{\omega_n^2-\omega^2}\cos\omega t$, represents the forced oscillation, and is a simple harmonic motion of the same period, $2\pi/\omega$, as the external disturbance.

We may note that:

(i) if $\omega_n > \omega$, the periodic time of the external disturbance, namely, $2\pi/\omega$ is greater than the periodic time of free oscillation, $2\pi/\omega_n$—i.e. the point P is moved relatively slowly in s.h.m;

(ii) if $\omega_n < \omega$, the periodic time of the forced (external) oscillation is less than that of the free oscillation or the frequency of the forced oscillation of P is greater than that of the mass m—i.e. P oscillates relatively quickly;

(iii) if $\omega_n = \omega$, the periodic times and frequencies are equal, and the above solution of the differential equation does not apply, for when $\omega_n = \omega$ the forced oscillation is $\dfrac{\omega_n}{2} at\sin\omega_n t$ which is a modified s.h.m. in which the amplitude is proportional to time, t, as in Fig. 4.1(g). As the value of x increases the original assumption that θ is small is violated, but the solution holds in the early stages of the motion and shows that the most vigorous forced oscillations occur when the external period is made equal to the given free period—this being the simplest case of resonance.

EXAMPLE 4.1(iii)—illustrative of the ratio of torsional amplitudes

A shaft keyed concentrically to a wheel undergoes harmonic torsional oscillation $\theta = \theta_0\sin\omega t$. Derive an expression for the ratio of the amplitude of the wheel to that of the shaft relative to (a) the shaft, and (b) a fixed line, assuming there is no damping.

Referring to Fig. 4.1(h) we see that $\phi = \theta - \alpha$. Hence, the torque exerted by the shaft on the wheel is $T = k\phi$, where $k = GJ/L$ (also see Fig. A.6(a)). Thus, $I\ddot{\alpha} = T = k\phi = I(\ddot{\theta}-\ddot{\phi})$ or $\ddot{\phi} + \omega_n\phi = \ddot{\theta}$, where $\omega_n = k/I$. Alternatively, using $\theta = \theta_0\sin\omega t$, we may write: $-\omega^2\phi_0 + \omega_n^2\phi_0 = -\omega^2\theta_0$.

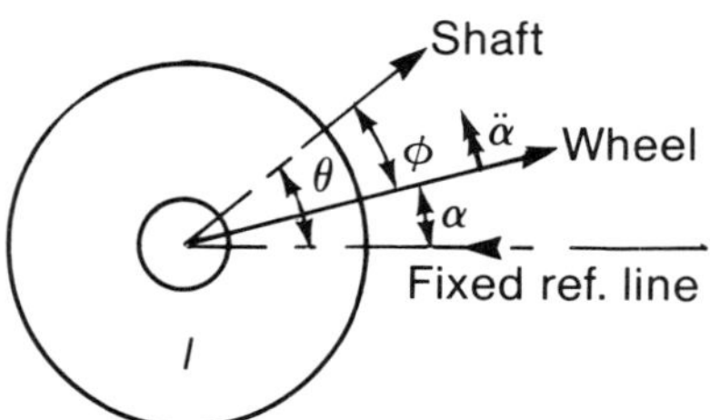

Fig. 4.1(h) Torsional oscillations

Hence, (a)

$$\frac{\phi_0}{\theta_0} = \frac{\omega^2}{\omega^2 - \omega_n^2} = \frac{\phi}{\theta} = -\left\{\frac{(\omega/\omega_n)^2}{1-(\omega/\omega_n)^2}\right\}$$

and (b)

$$\frac{\alpha}{\theta} = \frac{\omega_n^2}{\omega_n^2 - \omega^2} = \frac{1}{1-(\omega/\omega_n)^2} = 1 - \frac{\phi}{\theta}.$$

4.2 UNDAMPED VIBRATION-MEASURING INSTRUMENTS

An instrument recording displacement is often called a vibrometer; and an accelerometer is one which measures acceleration. Both can be understood with reference to Fig. 4.2(a), i.e. by considering the undamped induced oscillations of a spring-mass system with moving supports. (The case with damping is dealt with in Section 5.3).

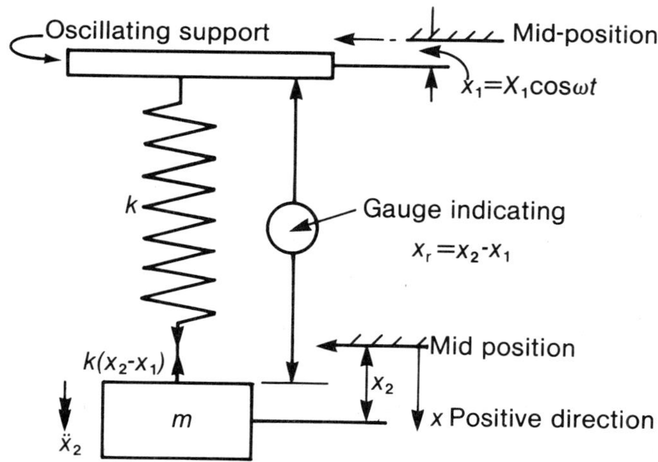

Fig. 4.2(a) Relative motion of seismic mass and moving support. (Note, $mg = kd$ and so does not appear on diagram or in equation of motion)

With negligible damping, the support (assumed oscillating with s.h.m. according to the law $x_1 = X_1 \cos \omega t$) will cause the free body of mass m to oscillate harmonically with the same frequency, i.e. $x_2 = X_2 \cos \omega t$, and the gauge will indicate the change in the length of the spring, namely, $x_r = x_2 - x_1$ which, in turn, exerts a force, $k(x_2 - x_1)$ on the mass m. Hence, the equation of motion of the mass is $m\dfrac{d^2x_2}{dt^2} + k(x_2 - x_1) = 0$ (see Section 2.2 re-omission of mg and kd) or $[D^2 + \omega_n^2]x_2 = kx_1$, where $\omega_n = \sqrt{(k/m)}$. Alternatively, $-m\omega^2 X_2 + kX_2 = kX_1$ since $\cos \omega t$ cancels throughout, and the phase diagrams are shown in Fig. 4.2(b). From the non-dimensional diagram we see that $\dfrac{X_1}{X_2} = 1 - \left(\dfrac{\omega}{\omega_n}\right)^2$. Hence, $X_2 - X_1 = X_1 \left\{\dfrac{(\omega/\omega_n)^2}{1 - (\omega/\omega_n)^2}\right\}$ which can be measured by means of a mechanical dial gauge of small inertia—the difference between the extreme scale-readings being $2(X_2 - X_1)$.

Fig. 4.2(c) clearly shows that when ω is low, $x_r = X_2 - X_1$ approaches zero, i.e. the system moves in unison; whereas if ω is high, $\dfrac{X_2 - X_1}{X_1}$ approaches negative unity, i.e. x_r approaches X_1 in the opposite direction achieved by use of a large mass and/or weak springs.

Thus, in the case of a vibrometer recording displacement or amplitude X_1 of an oscillating support, its accuracy depends on how low ω_n is relative to, ω, the excitation frequency. For example, seismographs recording earthquakes for which the periods may be relatively long, natural frequencies, ω_n, may be as low as 5 to 10 cycles/minute so that $\omega/\omega_n \gg 1$ and, hence, the recording of $x_r = X_2 - X_1$ approaches—X_1 which is the amplitude of the foundation or support frame. The size of m may be limited by the space into which it has to be suspended from the oscillating support whose amplitude, X_1, is being measured, e.g. for measuring the vibration of instrument-panels in aircraft, frameworks of buildings and for seismographs, etc.

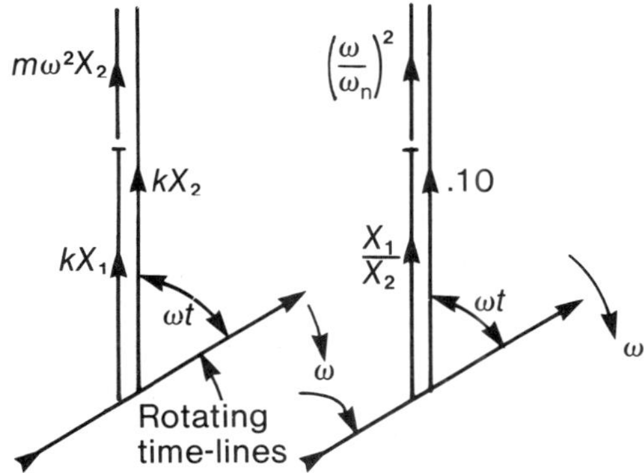

Fig. 4.2(b) Vector diagrams of forces and force ratios

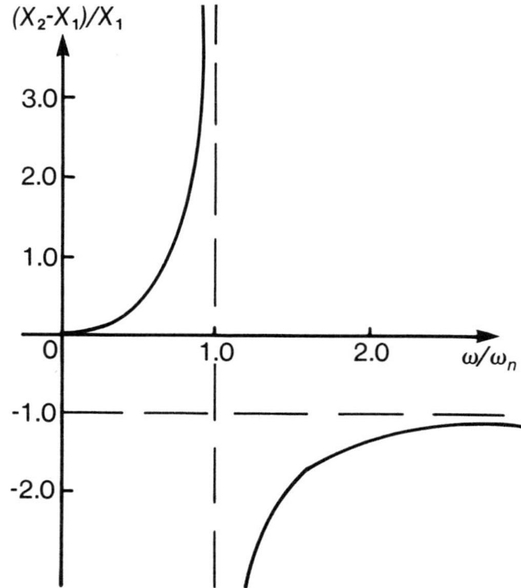

Fig. 4.2(c) Amplitude ratio against frequency ratio

Similarly, in the case of an accelerometer recording the acceleration ($\ddot{x}_1$) of a foundation or supporting frame, the natural frequency, ω_n (i.e. $\sqrt{(k/m)}$) must be high relative to the excitation frequency ω. Then, since $x_r = x_2 - x_1 = (X_2 - X_1)\cos \omega t = X_1 \left\{ \dfrac{(\omega/\omega_n)^2}{1 - (\omega/\omega_n)^2} \right\} \cos \omega t$,

$$x_r = \frac{X_1 \omega^2 \cos \omega t}{\omega_n^2\{1 - (\omega/\omega_n)^2\}} = \frac{\ddot{x}_1}{\omega_n^2\{(\omega/\omega_n)^2 - 1\}}.$$

If $(\omega/\omega_n) \ll 1$, x_r approaches proportionality with $\ddot{x}_1$ and the vibrational record of x_r will be a scale record of acceleration, $\ddot{x}_1$, of the base supporting the instrument. However, since ω_n is high, the presence of higher harmonics in resonance with the natural frequency would destroy the interpretation of an accelerometer record unless the instrument were damped (as in Section (5.3). In contrast, the displacement-measuring instrument or vibrometer which has a low natural frequency does not require damping. The output of an accelerometer if integrated twice results, of course, in a displacement-time record of vibration. We may summarize these cases of undamped vibration-measuring devices as follows:

(i) If ω/ω_n is very large the amplitude recorded is $-X_1$, i.e. the mass m remains almost stationary (as in seismographs) so that the amplitude of the relative motion (x_r) between the mass and the vibrating support is X_1, which is the amplitude of vibration required to be measured,

(ii) If ω/ω_n is very low the relative motion (x_r) between the mass and the vibrating support is equal to $\frac{1}{\omega_n^2}\times$ acceleration of the support required to be measured,

(iii) If $\omega/\omega_n = 1$ the relative motion (x_r) between the mass and the vibrating support becomes large; hence, an instrument capable of being tuned (e.g. the free length of a horizontal reed projecting from a block clamped to a vibrating support) so that it resonates when its natural frequency coincides with that of the vibrating support enables a vibration to be detected and its frequency to be measured.

Section 5.3 contains diagrams and analyses of systems which form the bases of damped vibration-measuring instruments, and Examples thereon involving quantitative work.

EXAMPLE 4.2(i)—illustrative of the use of a vibration pick-up

An undamped pick-up of the type shown diagrammatically in Fig. 4.2(a) has a natural frequency of 1 Hz and is mounted on a frame vibrating harmonically with a frequency of 5 Hz. If the relative amplitude between the pick-up mass and the vibrating frame is 2 mm, estimate the absolute amplitudes of the frame and mass.

The relative amplitude $X_2 - X_1 = X_1\left\{\frac{(\omega/\omega_n)^2}{1-(\omega/\omega_n)^2}\right\} = X_r$. Hence, the absolute amplitude of the frame is

$$X_1 = X_r\left\{\frac{1-(\omega/\omega_n)^2}{(\omega/\omega_n)^2}\right\} = 2\text{mm}\left\{\frac{1-25}{25}\right\} = -1.92\,\text{mm}$$

i.e. when X_r is 2 mm, X_1 is -1.92 mm and $X_2 = 0.08$ mm which means that when the absolute amplitude of the mass is 0.08 mm downwards the absolute amplitude of the frame is 1.92 mm upwards of its mid-position.

EXAMPLE 4.2(ii)—illustrative of the use of amplitude measurement

An instrument on the bridge of a ship is observed to vibrate with a certain amplitude when the propeller speed is 100 rev/min, and with eight times that amplitude when the speed is 200 rev/min. Assuming that the disturbing force varies as the square of the speed, estimate the speed at which resonance is probable.

$\frac{X}{F/k} = \frac{1}{1-\left(\frac{\omega}{\omega_n}\right)^2} = \frac{kX}{F}$ which is the amplitude ratio (see Fig. 4.1(b)) or the fraction of the maximum disturbing force, F, transmitted to supports, and

$$\frac{\omega}{\omega_n} = \frac{\text{frequency of disturbing force}}{\text{frequency of natural vibrations}} = \frac{\text{actual speed of propeller}}{\text{resonant speed of propeller}}$$

Let

$$r = \frac{100\,\text{rev/min}}{\text{resonant speed}}, \quad \text{then } \frac{X_{100}}{F/k} = \frac{1}{1-r^2}$$

and

$$\frac{X_{200}}{4F/k} = \frac{1}{1-4r^2} = \frac{8X_{100}}{4F/k} = 2\frac{X_{100}}{F/k} = \frac{2}{1-r^2}$$

Hence, $2 - 8r^2 = 1 - r^2$ or $7r^2 = 1$ and $r = 0.378$ from which it follows that the probable speed of the propeller at which resonance will occur is $\frac{100}{0.378}$ rev/min or 265 rev/min.

4.3 DAMPED FORCED OSCILLATIONS WITH ONE DEGREE OF FREEDOM

These concern elastic systems subjected to periodic impressed forces, and where energy is being dissipated by viscous damping. Such vibration problems may be investigated by considering analogous equivalent idealized systems. Two main categories of forced vibrations will be considered: (i) when the disturbing force is applied directly to the mass (as in Section 4.4), and (ii) when the base or frame-work of a spring-supported mass moves harmonically (as in Section 5.3).

We saw in Figs. 3.3(b) and (e) that the motion of an m, k, c system set going and left to itself would damp itself out. If it does not damp out, a periodic force must be being applied to keep it going, as illustrated in Fig. 4.4(a). A common source of periodically-varying impressed force occurs in reciprocating and rotary machinery—the cause being out-of-balance mass requiring a centripetal force ($F = mr\omega^2$) to keep it rotating about an axis which impresses a harmonic force ($f = F\cos\omega t$) in a linear direction on the framework of the system. To keep the transmitted vibration down to a minimum, the disturbing source must be isolated by flexible supports (steel springs, rubber, cork, etc) designed to suit the system (see Example 4.6(ii) and Fig. 4.6(b) for transmissibility curves).

Disturbances and forces transmitted to the mountings of machines or instruments, etc. cause vibratory responses (as in Section 5.3) and are easily predictable by means of phase-diagram analysis, and measurable by means of a (seismic) mass mounted on springs which represents the basis on which several types of vibration-measuring instruments are designed. The vibrational characteristics of structures are often explored by deliberately mounting on the structure two equal out-of-balance masses geared to rotate in opposite directions at the same speed, and the amplitude at resonance speed observed (see Example 4.4(v)). Alternatively, a simple iron-cored electromagnet placed near a magnetic component will excite this into a small forced vibration which will build up many-fold when a resonant frequency is reached (see Fig. 6.4(a)). There is no additional mass attached to the part to alter its frequency; hence this is a usual method of exciting steel parts.

4.4 DAMPED FORCED OSCILLATIONS ON RIGID SUPPORTS

Fig. 4.4(a) is diagrammatic of damped oscillatory systems acted on by harmonic forces. Mass m of weight mg stretches the spring by distance d such that $mg = kd$ before oscillations begin about the static-equilibrium or mid-position and are maintained by a periodic force $F\cos(\omega t + \beta)$ of maximum value F and denoted by f at any time t applied by some means—e.g. out-of-balance rotating parts (as in Fig. 4.4(k)). If the system is horizontal, the gravitational force does not appear in the equation of motion and, in any case (as in Section 2.2) mg cancels with kd in the equation referring to Fig. 4.4(a), namely:

$$mg + f - c\dot{x} - k(d + x) = m\ddot{x} = f - c\dot{x} - kx$$

i.e.

$$m\ddot{x} + c\dot{x} + kx = f = F\cos(\omega t + \beta) \tag{1}$$

The complete mathematical solution of this equation (see Appendix B11) is dealt with in Example 4.4(iii) and consists of a transient part corresponding

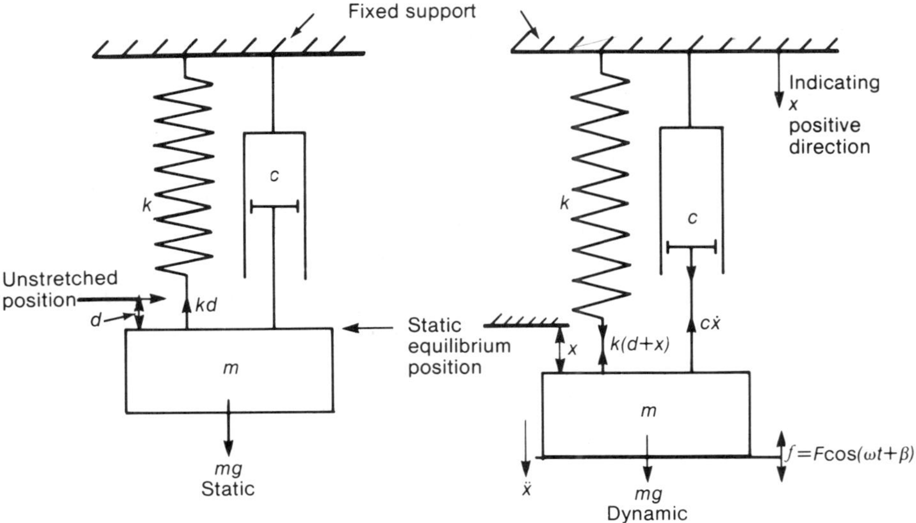

Fig. 4.4(a) Damped forced oscillatory system

to free vibration together with a constant amplitude or steady-state part which continues as long as the forcing function is active. When the transient has died away, the constant amplitude term of frequency ω can be interpreted (as follows) by rotating vectors which enable an understanding of the influence of mass, stiffness, damping and exciting frequency on amplitude and phase relationships to be obtained complementary to mathematical analysis. In particular, if the mass is forced to oscillate in s.h.m. (see Fig. 1.5(a)) as described by displacement $x = X\cos\omega t$, $\dot{x} = \omega X \cos\left(\omega t + \frac{\pi}{2}\right)$ or $v = V\cos\left(\omega t + \frac{\pi}{2}\right)$ and, $\ddot{x} = \omega^2 X\cos(\omega t + \pi)$ or $a = A\cos(\omega t + \pi)$ then equation (1) may be written as:

$$m\omega^2 X\cos(\omega t + \pi) + c\omega X\cos(\omega t + \pi/2) + kX\cos\omega t =$$
$$f = F\cos(\omega t + \beta), \tag{2}$$

These four terms may be represented graphically by vectors rotating relative to a time line (or vice versa as in Fig. 4.4(b)) which form a visual aid towards a clearer understanding of the action and dynamics of the system than can be obtained from a purely mathematical solution (see Example 4.4(iii)), Thus, after the transient vibrations have given way to steady-state vibrations the three terms on the left of equation (2) can be represented by three vectors spaced successively at right angles as in Fig. 4.4(b)—i.e. in the same way as was done in Fig. 2.3(a). The spring force at time t is kx (max value kX) and is always opposite in direction to the displacement x; the damping force cv (max. value $cV = c\omega X$) is opposite to velocity; the inertia force ma (max. value $mA = m\omega^2 X$) is opposite in direction to acceleration, and hence is in phase with displacement x since the phase differences between the maximum values of spring and damping forces is $\pi/2$, and of spring and inertia forces is π as shown on Fig. 4.4(b). It is, however, more useful to set the terms of equation (2) in the vector sequence shown in Fig. 4.4(c) in which OA represents the maximum value (kX) of the force exerted by the spring; AB, drawn at 90° counterclockwise, represents the maximum value ($c\omega X$) of the viscous damping force; BD represents the maximum value ($m\omega^2 X$) of the force accelerating the mass (m); and OD represents the maximum value (F) of the applied force which maintains the steady-state vibration. The instantaneous value of each force at any time t is represented by the projection (oa, ab, bd, od) of its maximum vector on the time line which rotates at a constant angular rate ω—the angular frequency of the forced oscillation (see also Figs. 2.3(a) and 3.3(d)). Since equations (1) and (2) require that, at every instant in the cycle, the sum of the three forces represented by the terms on the left (i.e. lengths oa, ab, bd) must vectorially be equal to the instantaneous

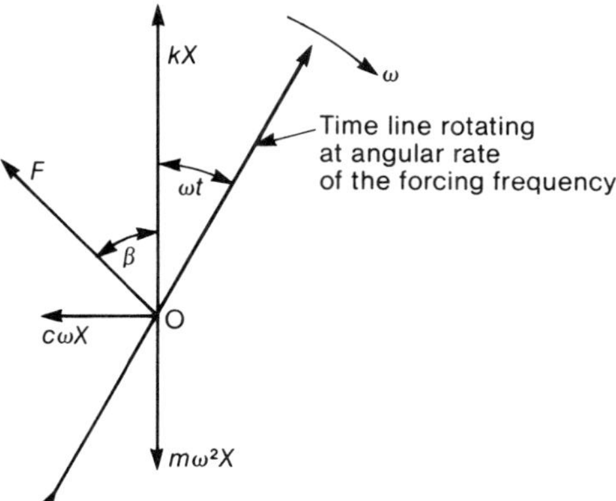

Fig. 4.4(b) Phase diagram of forces

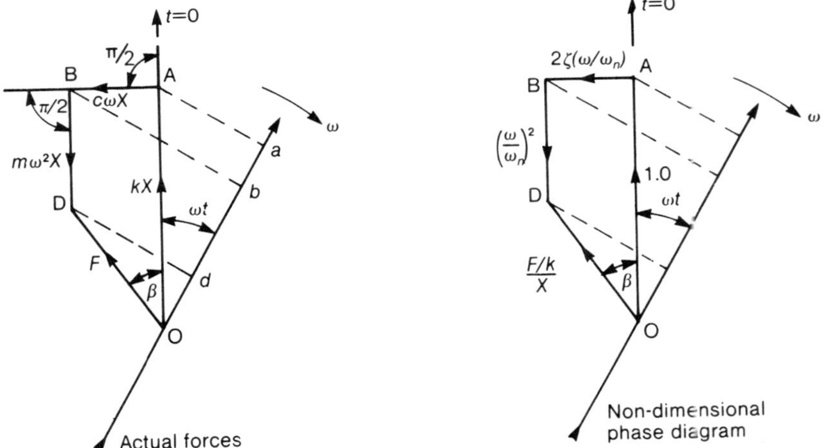

Fig. 4.4(c) Vector diagram of forces

value of the external force as represented by vector od, the vectors representing the maximum values (i.e. OA, AB, BD, OD) must form a closed figure—i.e. OD represents the vector sum of the other three and represents the maximum value, F, of the applied force. The instantaneous value of the applied force is $f = F\cos(\omega t + \beta)$, where angle AOD $= \beta$ is the phase lag between the maximum applied force vector F and the maximum displacement force vector kX.

Thus, we see from Fig. 4.4(c) that the equation of motion (1 or 2) expressed graphically shows how easily the ratio X/F and phase angle β can be estimated for known values of m, k, c and ω. The non-dimensional phase diagram

(Fig. 4.4(c)) showing force ratios (i.e. relative to the maximum spring force kX) enables previously deduced relationships (as in Section 3.2) to be used, namely : $\omega_n = \sqrt{(k/m)}$; $\zeta = c/c_c = \dfrac{c}{2\sqrt{(mk)}}$; $2\zeta\omega_n = c/m$; $2\zeta/\omega_n = c/k$.

The ratio F/kX is important—being the ratio between the maximum value of the applied harmonic force and the force which would be necessary to give the spring a *static* displacement equal to the maximum amplitude. The inverse of this, namely, $\dfrac{X}{F/k}$ is called the 'displacement ratio' or 'magnification ratio'—being the ratio of the amplitude of the harmonic motion and the displacement which would be produced *statically* by a steady force F applied to the spring of stiffness k—i.e. the ratio of amplitude/equivalent static displacement. When $\omega = \omega_n$, the phase diagrams of the damped forced system of Fig. 4.4(a) become rectangles as shown in Fig. 4.4(d)—the forcing frequency being equal to the undamped natural frequency. In the general case, we see from the non-dimensional diagrams of Figs. 4.4(c, e) that

$$\left(\frac{F}{kX}\right)^2 = \left\{1 - \left(\frac{\omega}{\omega_n}\right)^2\right\}^2 + \left\{2\zeta\left(\frac{\omega}{\omega_n}\right)\right\}^2.$$

Hence, the displacement ratio

$$\frac{X}{F/k} = \frac{1}{\sqrt{\left(\left\{1 - \left(\frac{\omega}{\omega_n}\right)^2\right\}^2 + \left\{2\zeta\left(\frac{\omega}{\omega_n}\right)\right\}^2\right)}} = \frac{X}{X_{st}} \qquad (3)$$

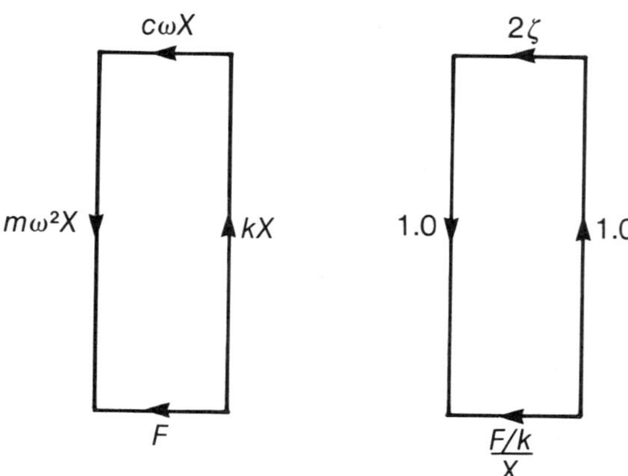

Fig. 4.4(d) Phase diagrams when $\omega = \omega_n$

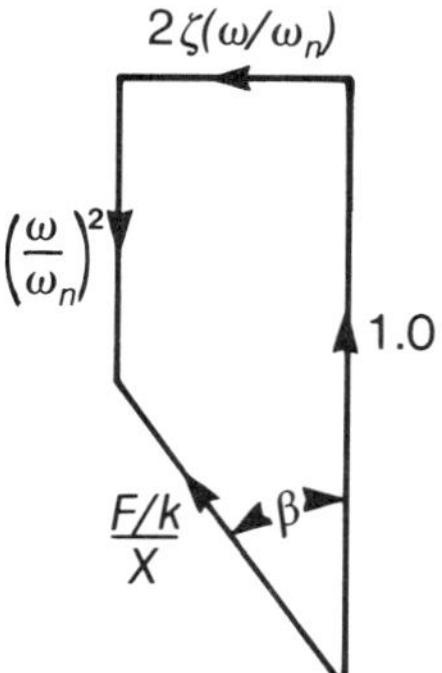

Fig. 4.4(e) General non-dimensional phase diagram

Also, the phase lag (angle β by which the response X lags behind the forcing function) is given by

$$\tan\beta = \frac{2\zeta(\omega/\omega_n)}{1-(\omega/\omega_n)^2} = \frac{2\zeta\omega\omega_n}{\omega_n^2-\omega^2} \tag{4}$$

We may note that dimensional analysis of damped forced oscillation systems with one degree of freedom would yield the non-dimensional parameters involved, namely, (ω/ω_n), ζ, and X/X_{st}. For, since the independent variables are x, m, k, c, f and its frequency ω, we may state that x is some function of the other five. The dimensionless groups or parameters (by the Π theorem) may be written:

$$\frac{x}{f/k} = \phi\left\{\left(\frac{\omega^2 m}{k}\right), \left(\frac{c}{2\sqrt{(mk)}}\right)\right\}$$

or

$$\frac{X}{X_{st}} = \phi\left\{\left(\frac{\omega}{\omega_n}\right)^2, \zeta\right\},$$

where the function ϕ has to be determined. This is easily obtained by analysis which results in equation (3) previous.

Graphs of equations (3) and (4) (which can be plotted on a micro-computer provided it has high resolution graphics) show the curves of Figs. 4.4(f) and 4.4(g) for various amounts of viscous damping expressed by the value of the damping factor ζ. From these graphs we see that:

(i) at low frequency ratios the displacement ratio is nearly unity,

(ii) when the frequency ratio is 1.0—i.e. the forcing frequency (ω) of the applied force is equal to the natural frequency (ω_n) of the system, the displace-

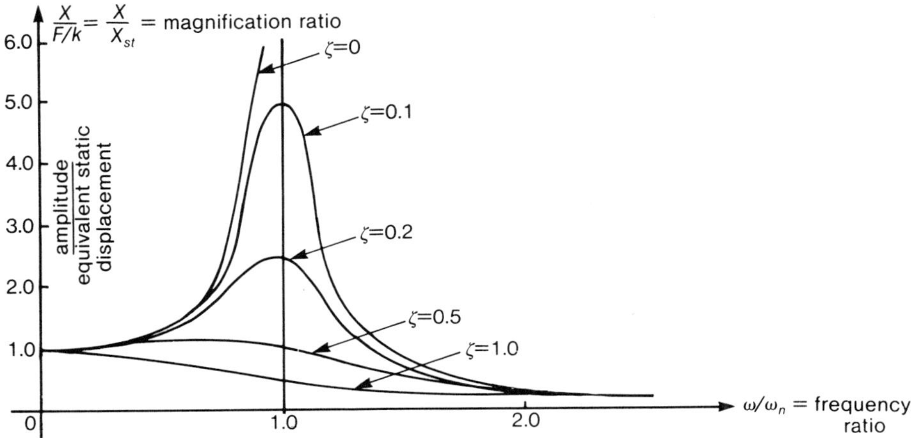

Fig. 4.4(f) Effect of damping factor on displacement ratio

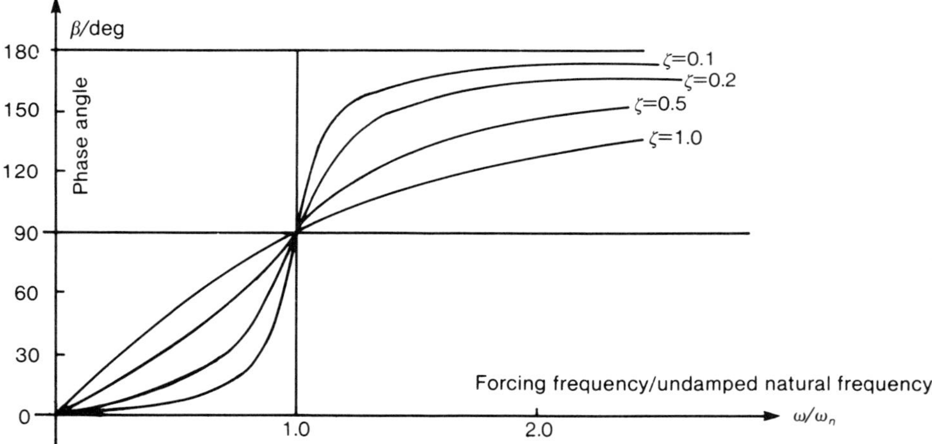

Fig. 4.4(g) Phase lag of response behind forcing function

ment ratio $\frac{X}{F/k} = \frac{1}{2\zeta}$ (see Fig. 4.4(d)) showing that unless the damping factor is high, the amplitude, X, of the vibration will be relatively high.

(iii) as the frequency ratio increases beyond resonance (i.e. the frequency of maximum response or amplitude), the amplitude of vibration falls rapidly, and the displacement ratio tends to zero,

(iv) the curves come down steeper after maximum amplitude than when rising to it—i.e. the condition of resonance or critical speed is found more easily when speeding up ω from zero than when slowing down from a speed above the critical,

(v) with low-damping factors, ζ, (i.e. "light" damping) the phase lag angle, β, remains small until $\omega/\omega_n = 1$ and $\beta = \pi/2$ or 90° after which β approaches π or 180° at higher frequency ratios—i.e. at high frequency ratios the applied force, f, is almost opposite in phase to the displacement for the reason that, when $\omega \gg \omega_n$, the force is almost wholly used to cause acceleration (i.e. overcoming the inertia force) of the mass, and directed towards the mid-position of the oscillation in the case of s.h.m. Thus, f attains almost its maximum value, F, when the displacement reaches maximum value, X, and velocity $v = V\cos\left(\omega t + \dfrac{\pi}{2}\right)$ is zero.

The phase diagrams for the range of frequency ratios below and greater than 1.0 (i.e. $\omega \langle = \rangle \omega_n$) for forced vibrations of an elastic system with very light damping (i.e. ζ is small) are shown in Fig. 4.4(h). At resonance (i.e. the frequency of maximum response) the peak amplitude of motion becomes very large if damping is 'light' (see Fig. 4.4(f)), and a spring-mass system literally tears itself apart. Variable-speed mechanical shakers (illustrated diagrammatically in Fig. 4.4(k)) in which the frequency, ω, of the harmonic force can be gradually increased have many industrial applications especially regarding the

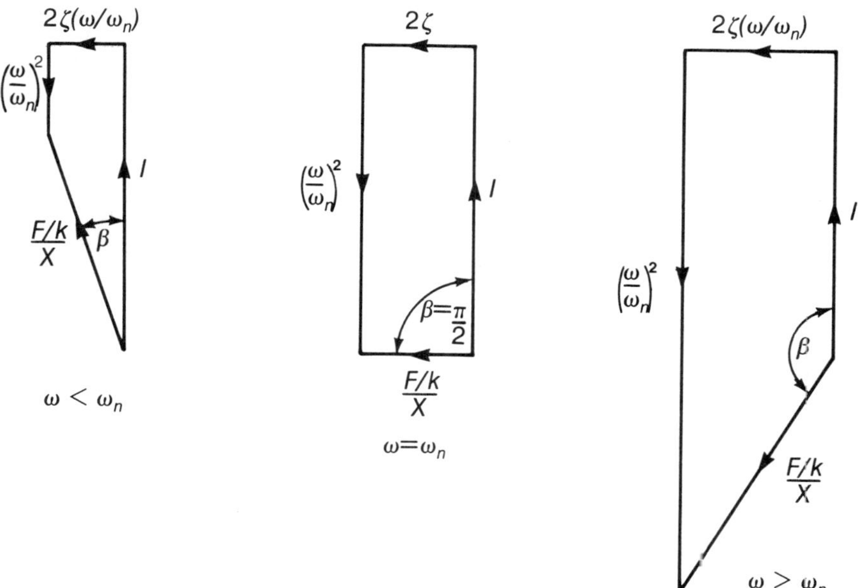

Fig. 4.4(h) Phase diagrams of damped forced oscillations

question of how far from the frequency causing maximum amplitude or the resonant condition represents safe operation.

A system in resonance (ω equal to or near to ω_n when damping is small) is obviously in a critical state, for the large displacements involved induce correspondingly high stresses in materials. In most cases, therefore, resonance must be avoided, or damage will result. From Figs. 4.4(d) and (f) it will be seen that, when damping is small, resonance occurs when $\omega/\omega_n = 1$ i.e. inertia force $m\omega^2 X$ is balanced by the spring force kX, and the applied force F overcomes the damping force $c\omega X$. The amplitude is $X = F/c\omega = F/c\omega_n$ or $F/2\zeta k$ and is relatively large (see Fig. 4.4(f)). The question of how far from resonance represents safe operation is important and can be answered by considering the force transmitted to supports (see Section 4.6). The maximum value of the displacement ratio $\left(\dfrac{X}{F/k}\right)$ or magnification ratio can be found from equation (3) as occurring when $(\omega/\omega_n) = \sqrt{(1 - 2\zeta^2)}$.

From this we may note that peak amplitude can (theoretically) occur at zero frequency when $\zeta^2 = 0.5$ or $\zeta = 0.707$ and diminishes from this peak with increasing frequency ω. Also, we may note, since $\omega_d = \omega_n\sqrt{(1 - \zeta^2)}$ for damped natural oscillations (as in Section 3.3), that for damped oscillations peak amplitudes or resonance states, occurring when $\omega = \omega_n\sqrt{(1 - 2\zeta^2)}$, do not occur at the frequency of damped natural oscillations but at a slightly less frequency (see Example 4.4(i)). The value of peak amplitudes of damped forced oscillations is given by $\left(\dfrac{X}{F/k}\right)_{max} = \dfrac{1}{2\zeta\sqrt{(1 - \zeta^2)}}$ (see Fig. 4.4(f)). The phase lag is given by equation (4) which, for peak amplitudes, becomes $\tan\beta = \dfrac{2\zeta\sqrt{(1 - 2\zeta^2)}}{2\zeta^2} = \dfrac{1}{\zeta}\sqrt{(1 - 2\zeta^2)}$. A common problem concerns the minimizing or prevention of out-of-balance forces of rotary machinery being transmitted to foundations or surrounding structures (see Section 4.6). This isolation may be accomplished by mounting such machines on flexible supports such as rubber pads or easily-deformable material comprising a damping factor as well as a spring strength (see Fig. 4.4(k) & Example 4.6(ii)).

EXAMPLE 4.4(i)—illustrative of the effect of damping coefficient on resonant amplitude

An instrument consists of a mass on a spring of stiffness 1.75 N/mm, and a dashpot providing viscous damping. After an initial displacement on a scale recording 100 divisions from the equilibrium position, the successive amplitudes are -20, $+4$, -0.8 when the period of damped oscillation is 1 s. Estimate the damping coefficient, the equivalent mass and the frequency of the applied force to give maximum amplitude.

When this m, k, c system is disturbed and then left to itself, the ratio of the amplitudes of successive cycles is $\dfrac{100}{4} = \dfrac{-20}{-0.8} = 25$ and, hence, the logarithmic decrement is

$$\log_\varepsilon 25 = \frac{2\pi\zeta}{\sqrt{(1-\zeta^2)}} = 3.219, \quad \text{i.e. } \zeta = 0.455$$

Also, since $\tau_d = 1\,\text{s} = \dfrac{1}{\text{Hz}}$, the frequency of the natural damped oscillation is $1\,\text{Hz}\left[\dfrac{2\pi\,\text{rad}}{\text{Hz s}}\right] = 2\pi\,\text{rad/s} = \omega_d$ and the frequency of natural damped oscillations would be $\omega_n = \dfrac{\omega_d}{\sqrt{(1-\zeta^2)}} = \dfrac{2\pi}{\sqrt{(1-0.2075)}} = 7.06$ rad/s. Hence, the equivalent mass is

$$m = \frac{k}{\omega_n^2} = \frac{1.75\times 10^3}{50\,\text{rad}^2}\frac{\text{Ns}^2}{\text{m}}\left[\frac{\text{kg m}}{\text{Ns}^2}\right] = 35\,\text{kg}.$$

For maximum amplitude, $\dfrac{\omega}{\omega_n} = \sqrt{(1-2\zeta^2)} = 0.764$ and $\omega_n = 5.4\,\dfrac{\text{rad}}{\text{s}}\left[\dfrac{\text{Hz s}}{2\pi\,\text{rad}}\right] =$ 0.86 Hz.

EXAMPLE 4.4(ii)—illustrative of resonant frequency and amplitude of the mountings of a machine

A machine of mass 90 kg is supported on springs of total stiffness 36 N/mm. If a harmonic disturbing force $f = F\cos(\omega t + \beta)$ of maximum value 45 N acts on the machine, determine the resonant frequency and resonant amplitude assuming a viscous damping coefficient $c = 1$ Ns/mm. Also, estimate the angular frequency for peak amplitude and its phase angle.

The system is fundamentally that illustrated in Fig. 4.4(a) and the general phase diagram is that shown in Fig. 4.4(c). In this particular case, the natural frequency is

$$\omega_n = \sqrt{(k/m)} = \sqrt{\left(\frac{36\,\text{N}}{\text{mm}} \times \frac{1}{90\,\text{kg}}\left[\frac{\text{kg}\,1000\,\text{mm}}{\text{N s}^2}\right]\right)} = 20/\text{s} = 3.18\,\text{cycle/s}.$$

The peak amplitude or resonance occurs when $\left(\dfrac{X}{F/k}\right)_{max} = \dfrac{1}{2\zeta\sqrt{(1-\zeta^2)}}$ where

$$\zeta = \frac{c}{2\sqrt{(mk)}} = \frac{1\,\text{Ns/mm}}{2\sqrt{\left(90\,\text{kg}\times 36\dfrac{\text{N}}{\text{mm}}\left[\dfrac{\text{N s}^2}{\text{kg}\,10^3\,\text{mm}}\right]\right)}} = \frac{10}{2\sqrt{(9\times 36)}} = \frac{5}{18}$$

$$= 0.278$$

Hence,

$$X_{max} = \frac{45\,\text{N mm}}{36\,\text{N}} \times \frac{1}{0.556\sqrt{0.9227}} = 2.34\,\text{mm}$$

$$\text{and } \tan\beta_{peak} = \frac{\sqrt{(1-2\zeta^2)}}{\zeta} = \frac{\sqrt{0.8454}}{0.278} = 3.31, \text{ i.e. } \beta_{peak} = 73^\circ 12'$$

Also, the resonant frequency,

$$\omega_r = \omega_n\sqrt{(1-2\zeta^2)} = \frac{20}{s}\sqrt{0.8454} = 18.4/s\left[\frac{\text{Hz s}}{2\pi}\right] = 2.93\ \text{Hz}$$

i.e. resonance occurs when $\omega/\omega_n = 0.92$.

EXAMPLE 4.4(iii)—illustrative of a mathematical rather than a graphical solution
Consider the equations resulting from the analysis of the damped forced oscillations represented by Fig. 4.4(a) (i.e. an m, k, c, f system) in which the impressed frequency is equal to the undamped resonant frequency (i.e. $\omega = \omega_n$) and the initial displacement and velocity are zero (i.e. when $t = 0$, $x = 0$ and $\dot{x} = 0$). Deduce an equation which applies to data of Example 4.4(ii).

Using $\omega_d = \omega_n\sqrt{(1-\zeta^2)}$; $\omega_n^2 = k/m$ and $2\zeta\omega_n = c/m$ as before (see Sections 3.2 and 3.4), equation (1) of Section 4.4 may be written $[D^2 + 2\zeta\omega_n D + \omega_n^2]\,x = f/m$ and the mathematical solution (see Appendix B13) is

$$x = C\varepsilon^{-\zeta\omega_n t}\cos(\omega_d t + \alpha) + \frac{F}{m}\frac{\cos\omega t}{\sqrt{\{(\omega_n^2-\omega^2)+(2\zeta\omega_n\omega)^2\}}}$$

the first term of which is transient and dies away in time depending on initial conditions or how the motion is started, whereas the second term (the steady-state term) is independent of the initial conditions. The velocity of the mass at any instant is given by

$$\dot{x} = C\varepsilon^{-\zeta\omega_n t}\{-\zeta\omega_n\cos(\omega_d t+\alpha) - \omega_d\sin(\omega_d t+\alpha)\} - \frac{F}{m}\omega\frac{\sin\omega t}{\sqrt{\{(\omega_n^2-\omega^2)+(2\zeta\omega_n\omega)^2\}}}$$

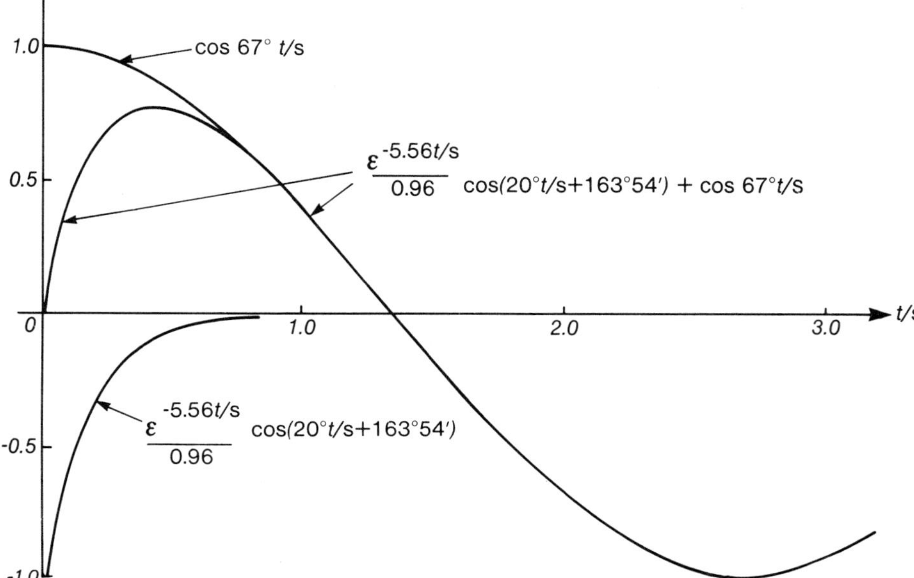

Fig. 4.4(i) Effect of transient term

Hence, since $\dot{x}=0$ when $t=0$, we have $0=-C\times 1\{-\zeta\omega_n\cos\alpha-\omega_d\sin\alpha\}-0$, i.e. $\tan\alpha=-\dfrac{\zeta\omega_n}{\omega_d}=\dfrac{-\zeta}{\sqrt{1-\zeta^2}}$ from which we may take $\sin\alpha=\zeta$ and $\cos\alpha=-\sqrt{(1-\zeta^2)}$. Also, since $x=0$ when $t=0$, we have, for the resonant case when $\omega=\omega_n$, $0=C\times 1\cos\alpha+\dfrac{F}{m2\zeta\omega_n\omega}=C\cos\alpha+\dfrac{F}{c\omega_n}$ i.e. $C=\dfrac{F}{c\omega_n\sqrt{(1-\zeta^2)}}$. Thus, the final solution satisfying the initial conditions of $x=0$ and $\dot{x}=0$ when $t=0$, and $\omega=\omega_n$ is $x=\dfrac{F}{c\omega_n}\left\{\dfrac{\varepsilon^{-\zeta\omega_n t}}{\sqrt{(1-\zeta^2)}}\cos(\sqrt{(1-\zeta^2)}\omega_n t+\sin^{-1}\zeta)+\cos\omega_n t\right\}$ which represents an oscillation that starts with zero amplitude and zero velocity, and builds up to a steady oscillation represented by $x=\dfrac{F\cos\omega_n t}{c\omega_n}$ or $\dfrac{F\cos\omega_n t}{2\zeta k}$. Thus, taking data of Example 4.4(ii) namely, $\omega=\omega_n=20/\mathrm{s}$; $\zeta=0.278$; $F=45\mathrm{N}$; $k=36$ N/mm; $c=1$ Ns/mm, the displacement equation becomes $x=2.25\,\mathrm{mm}\left\{\dfrac{\varepsilon^{-5.56t/s}}{0.96}\cos\left(20°\dfrac{t}{\mathrm{s}}+163°54'\right)+\cos 67°\dfrac{t}{\mathrm{s}}\right\}$.

The effect of the transient is to prevent the displacement at the start having the same simple harmonic variation as the applied force. This initial distortion of wave-shape can be troublesome, e.g. in servomechanisms. Fig. 4.4(i) shows a plot of the two terms within the brackets, and from which it can be seen that the transient term dies out within the first second of time and that the oscillation quickly gets into the steady-state, as given by the second term. The ordinates of Fig. 4.4(i) when multiplied by 2.25 given the displacement, x, of the mass at any instant.

EXAMPLE 4.4(iv)—illustrative of the estimation of resonant frequency and damping in torsional vibrations

A torsional system with viscous damping has a moment of inertia $I=2\,\mathrm{Nms}^2$ and torsional stiffness $k=110\,\mathrm{kN\,m/rad}$. When subjected to a harmonic torque of maximum value 9 Nm, the resonant amplitude was found to be 0.009 degree. Estimate the damping factor, the resonant frequency, and the coefficient of viscous damping.

For a torsional system, equation (1) becomes

$$I\ddot{\theta}+c\dot{\theta}+k\theta=T\cos(\omega t+\beta),\ \text{where}\ \theta=\Theta\cos\omega t.$$

Alternatively, $[\mathrm{D}^2+2\zeta\omega_n\mathrm{D}+\omega_n^2]\theta=\dfrac{T}{I}\cos(\omega t+\beta)$, where $\omega_n=\sqrt{(k/I)}$ and $2\zeta\omega_n=c/I$. Hence, the undamped natural frequency is $\omega_n=\sqrt{\left(\dfrac{110\,\mathrm{kNm}}{2\,\mathrm{Nms}^2}\right)}=235$ rad/s $=37.4$ cycle/s. The 'displacement ratio' analogous to equation (3) is $\dfrac{\Theta}{T/k}=\dfrac{1}{\sqrt{\left(\left\{1-\left(\dfrac{\omega}{\omega_n}\right)^2\right\}^2+\left\{2\zeta\dfrac{\omega}{\omega_n}\right\}^2\right)}}$ as deduced from the general non-dimensional phase diagram shown in Fig. 4.4(j). Peak amplitude (i.e. resonance) occurs at the resonant frequency $\omega_r=\omega_n\sqrt{(1-2\zeta^2)}$, and is given by $\dfrac{\Theta}{T/k}=\dfrac{1}{2\zeta\sqrt{(1-\zeta^2)}}$.

Hence, $\zeta\sqrt{(1-\zeta^2)}=\dfrac{9\,\mathrm{Nm}}{220\,\mathrm{kNm}\times 0.009\,\mathrm{deg}}\left[\dfrac{180\,\mathrm{deg}}{\pi}\right]$ and $\zeta=0.282$.

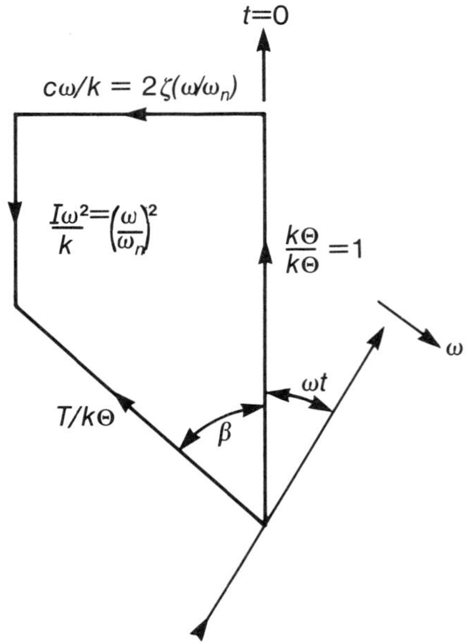

Fig. 4.4(j) Non-dimensional phase diagram for torsional oscillations

The resonant frequency $\omega_r = 37.4$ cycle/s $\sqrt{(1-0.16)} = 34.3$ cycle/s and the damping coefficient $c = 2\zeta\omega_n I = 2 \times 0.282 \times \dfrac{235}{\text{s}} \times 2\ \text{Nms}^2 = 265$ Nms or $\dfrac{\text{Nm}}{\text{rad/s}}$.

EXAMPLE 4.4(v)—illustrative of the use of an exciter and use of a stroboscope
A counter-rotating eccentric-mass exiciter with imbalance in each wheel of 45 kg mm is used to determine the vibrational characteristics of a structure of mass 180 kg. A stroboscope showed that the structure was passing upwards through its static-equilibrium position with the eccentric masses at the top when their speed was 15 rev/s, the amplitude being 25 mm. Estimate (a) the natural frequency and the damping factor of the structure, (b) the amplitude when the speed of the exciter is increased to 20 rev/s, and (c) the phase angle β—i.e. the angular position of the eccentric masses at the instant the strucure is moving upward through its equilibrium position.

The system is shown diagrammatically in Fig. 4.4(k) (a) The stroboscope showed the resonant state, i.e. F is a maximum when $x = 0$ and $\beta = 90°$ at a frequency of 15 cycles/s or angular speed $\omega = \omega_n = 30\pi$ rad/s = 15 Hz. From the phase diagram for $\omega = \omega_n$ which, for light damping, is very near to the state of resonance (Fig. 4.4(d)) we see that $\dfrac{F}{kX} = 2\zeta$. Hence the damping factor

$$\zeta = \frac{F}{2kX} = \frac{2m^1 r\omega_n^2}{2m\omega_n^2 X} = \frac{45\,\text{kg mm}}{180\,\text{kg}\,25\,\text{mm}} = \frac{1}{100} = 0.01$$

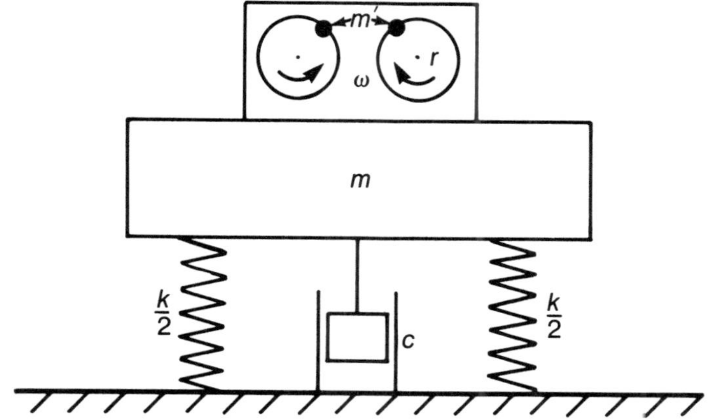

Fig. 4.4(k) Counter-rotating escentric-mass exciter

(b) If the speed is increased such that $\omega/\omega_n = 4/3$ the phase diagram becomes similar to that shown in Fig. 4.4(h) where $\omega > \omega_n$ and from which we deduce that

$$\frac{X}{F/k} = \frac{1}{\sqrt{\left(\left\{1-\left(\frac{\omega}{\omega_n}\right)^2\right\}^2 + \left\{2\zeta\left(\frac{\omega}{\omega_n}\right)\right\}^2\right)}} = \frac{1}{\sqrt{\left(\left\{1-\frac{16}{9}\right\}^2 + \left\{\frac{2}{100}\times\frac{16}{9}\right\}^2\right)}} = \frac{9}{7}$$

Hence the amplitude $X = \frac{9}{7} \times \frac{2m^1 r\omega^2}{m\omega_n^2} = \frac{18}{7} \times \frac{45\,\text{kg mm}}{180\,\text{kg}} \times \frac{16}{9} = \frac{16}{14}\,\text{mm}$ and

$$\tan\beta = \frac{2\zeta(\omega/\omega_n)}{1-(\omega/\omega_n)^2} = \frac{2}{100} \times \frac{4}{3} \times \left(\frac{-9}{7}\right) = -0.0343 \text{ i.e. } \beta = 180^\circ - 2^\circ = 178^\circ.$$

EXAMPLE 4.4(vi)—illustrative of information deducible from a vibration wave

A mass attached to a spring of stiffness 0.5 N/mm has a viscous damping device fitted. When the mass was displaced and released, the period of the vibration was observed to be 1.8 s and the ratio of consecutive amplitudes was estimated from a trace as 4.2. Determine the logarithmic decrement, the damping factor and the natural frequency of the system. Also, determine the amplitude and phase angle when the harmonic force of $9\cos 3t/\text{s}$ newtons acts on the system.

Referring to Fig. 3.6(c) we have $\tau_d = \frac{2\pi}{\omega_d} = 1.8\,\text{s}$ and $x_1/x_2 = 4.2$

Hence, logarithmic decrement $\delta = \log_\varepsilon(x_1/x_2) = \zeta\omega_n\tau_d = \frac{2\pi\zeta}{\sqrt{(1-\zeta^2)}}$ is

$$\delta = \log_\varepsilon 4.2 = 1.435 \quad \text{and}$$

$$\zeta^2 = \frac{\delta^2}{4\pi^2 + \delta^2} = \frac{2.06}{41.54} = 0.0495, \quad \text{i.e. } \zeta = 0.222$$

and

$$\omega_n = \frac{\delta}{\zeta\tau_d} = \frac{1.435}{0.399\,\text{s}} = 3.59/\text{s} \quad \text{or} \quad \frac{3.59}{\text{s}}\left[\frac{\text{Hz s}}{2\pi}\right] = 0.572\,\text{Hz}$$

If the system is forced to vibrate by an impressed harmonic force then, referring to Fig. 4.4(e), we see that

$$\frac{Xk}{F} = \frac{1}{\sqrt{\left(\{1-(\omega/\omega_n)^2\}^2 + \left\{2\zeta\frac{\omega}{\omega_n}\right\}^2\right)}} \quad \text{and } \tan\beta = \frac{2\zeta\omega/\omega_n}{1-(\omega/\omega_n)^2}$$

Hence,

$$X = \frac{9\,\text{N mm}}{0.5\,\text{N}} \frac{1}{\sqrt{\left(\left\{1-\left(\frac{3}{3.59}\right)^2\right\}^2 + \left\{2\times 0.222 \times \frac{3}{3.59}\right\}^2\right)}} = 2.71\,\text{mm}$$

and

$$\tan\beta = \frac{2\times 0.222\times 0.836}{1-0.7} = 1.237, \quad \text{i.e. } \beta = 51°3'.$$

4.5 LABORATORY WORK ON FORCED VIBRATIONS

Object To obtain the damping factor and resonance curve of a massive spring-controlled system with one degree of freedom, assuming damping proportional to velocity.

Apparatus and method Referring to apparatus shown diagrammatically in Fig. 4.5(a). First estimate the periodic time, τ_d, and the angular frequency, ω_d, of damped natural oscillation by switching off the motors and letting go of the statically-extended spring so that the system oscillates naturally. Record the 'signature' or decay of oscillations on a trace paper so that the logarithmic decrement $\delta = \log_\varepsilon\left(\frac{x_1}{x_2}\right) = \frac{1}{N}\log_\varepsilon\left(\frac{x_1}{x_{N+1}}\right)$ can be determined. Also, record the time for a counted number of oscillations to determine the period time, τ_d. Hence, using the formulae developed in Section 3.6 and used in Example 4.4(vi), determine the damping factor $\zeta = \frac{\delta}{\sqrt{(4\pi^2+\delta^2)}}$ and $\omega_n = \frac{\delta}{\zeta\tau_d}$ which can be checked with $\sqrt{(k/m)}$.

Secondly, with the motors running at any angular speed, ω, the system is forced to oscillate because of the vertical impressed force $f = F\cos\omega t$, where $F = 2m^1\gamma\omega^2$ and m^1 is the mass of each pin or bolt causing imbalance in the contra-rotating variable-speed motors.

Observations and results Observe and record the amplitude X and, by recording the time for, say, 10 cycles, deduce ω and F. Hence, for a

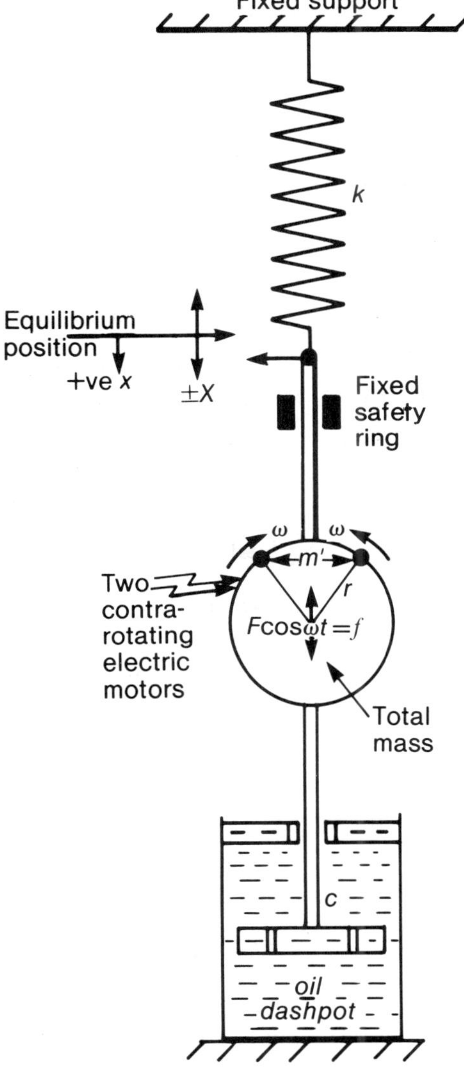

Fig. 4.5(a) Forced oscillations

series of speeds, tabulate ω, X, $\dfrac{X}{F/k}$, ω/ω_n and plot the non-dimensional curve with $\dfrac{X}{F/k}$ as ordinate and ω/ω_n as abcissa. Similarly $\beta = \tan^{-1}\left(\dfrac{2\zeta(\omega/\omega_n)}{1-(\omega/\omega_n)^2}\right)$. Also, plot the theoretical curves of

$$\frac{X}{F/k} = \frac{1}{\sqrt{(\{1-(\omega/\omega_n)^2\}^2 + (2\zeta\omega/\omega_n)^2)}}$$

and

$$\beta = \tan^{-1}\left(\frac{2\zeta(\omega/\omega_n)}{1-(\omega/\omega_n)^2}\right)$$

against ω/ω_n so as to compare the experimental results with the theoretical predictions, and draw conclusions.

EXAMPLE 4.5(i)—illustrative of information deducible by measurement of decaying amplitudes

A mass suspended on an elastic spring and controlled by a dashpot is observed to oscillate with a frequency of 3 cycles/s when disturbed, and to lose two thirds of its amplitude in 5 oscillations.

If the same system has a force $f = F\cos(pt + \beta)$ applied to it where $p = 2$ Hz and the steady amplitude is observed to be 3.3 mm, estimate the value of F and β assuming that the damping force exerted by the dashpot is given by $\frac{u}{7}\frac{\text{N}}{\text{mm/s}}$, where u represents its velocity.

Logarithmic decrement $\delta = \frac{1}{N}\log_\varepsilon\left(\frac{x_1}{x_{N+1}}\right) = \zeta\omega_n\tau_d = \frac{2\pi\zeta}{\sqrt{(1-\zeta^2)}}$. Hence, $\frac{1}{5}\log_\varepsilon 3 = \zeta\omega_n\ \text{s}/3 = \frac{2\pi\zeta}{\sqrt{(1-\zeta^2)}} = \frac{1.0986}{5} = 0.21972$ i.e. $\zeta = 0.035$ and $\omega_n = 18.84/\text{s}$.

$p = 2\,\text{Hz}\left[\frac{2\pi}{\text{Hzs}}\right] = 4\pi/\text{s} = \omega$ in the phase diagram of Fig. 4.5(b) and $c = \frac{\text{damping force}}{\text{velocity}} = \frac{1}{7}\frac{\text{N}}{\text{min/s}}$. Also, $\omega/\omega_n = \frac{4\pi}{18.84} = \frac{2}{3}$.

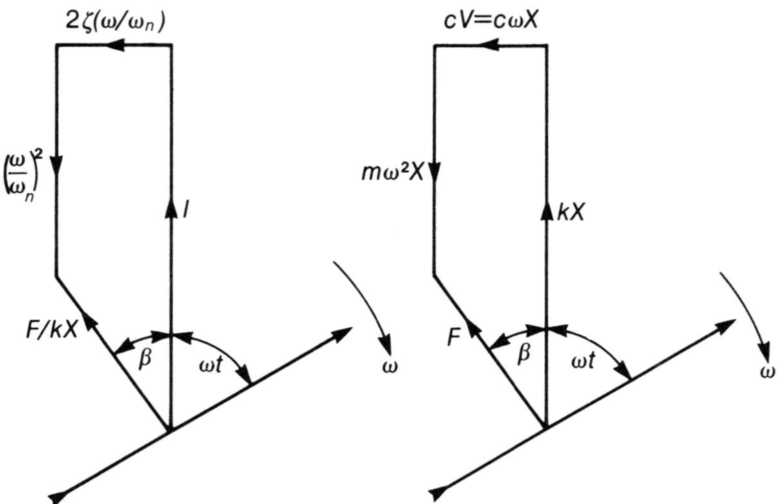

Fig. 4.5(a) Phase diagrams

$$\tan\beta = \frac{2\zeta(\omega/\omega_n)}{1-(\omega/\omega_n)^2}$$

$$= \frac{2\times 0.035\times(\frac{2}{3})}{1-\frac{4}{9}} = 0.084$$

$$\beta = 4°48'.$$

Hence,

$$\sin\beta = \frac{q\omega X}{F} \quad \text{i.e. } F = \frac{c\omega X}{\sin\beta} = \frac{\text{Ns}}{7\,\text{mm}} \times \frac{18.84}{\text{s}} \times \frac{3.3\,\text{mm}}{0.0837} = 106.3\,\text{N}$$

and $f = 106.3$ N $\cos(4\pi t/s + 4°48')$.

4.6 TRANSMISSIBILITY AND VIBRATION ISOLATION

Referring to Fig. 4.4(k) the harmonic force originating in the supported mass can only reach the foundation through the springs and dashpot. The instantaneous force (f_t) transmitted is the vectorial sum of the damping and spring forces as shown in Fig. 4.6(*a*), and the maximum force transmitted to foundations is

$$F_t = \sqrt{\{(kX)^2 + (c\omega X)^6\}} = kX\sqrt{\left\{1+\left(\frac{2\omega}{\omega_n}\zeta\right)^2\right\}}$$

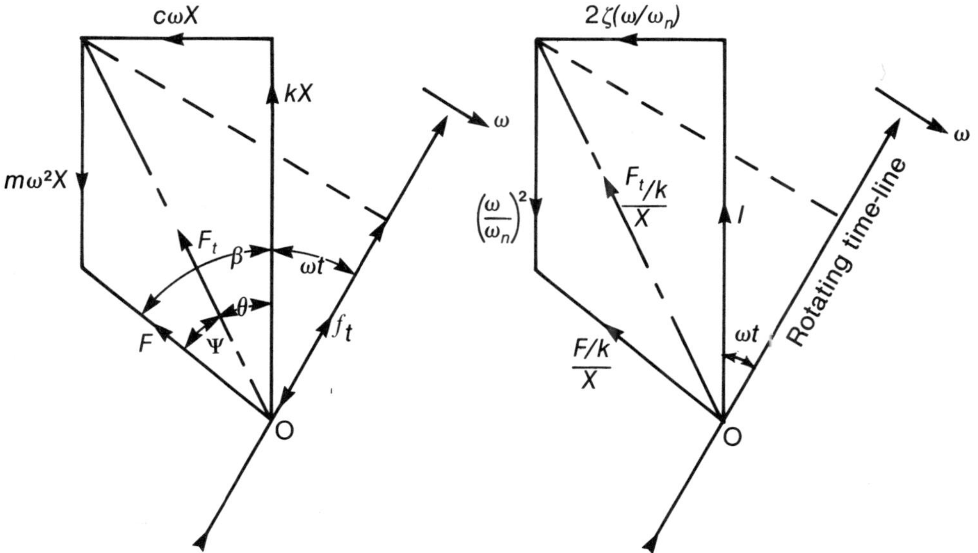

Fig. 4.6(a) Force transmitted to foundations

Also, since

$$\frac{kX}{F}=\frac{1}{\sqrt{\left(\left\{1-\left(\frac{\omega}{\omega_n}\right)^2\right\}^2+\left\{2\zeta\frac{\omega}{\omega_n}\right\}^2\right)}},$$

the ratio

$$\frac{F_t}{F}=\sqrt{\left(\frac{1+(2\omega/\omega_n\zeta)^2}{\left\{1-\left(\frac{\omega}{\omega_n}\right)^2\right\}^2+\left\{2\zeta\frac{\omega}{\omega_n}\right\}^2}\right)}$$

where F is the maximum value of the impressed force—i.e. of the force which would be transmitted to the foundation if the system were rigidly fixed to it without springs.

Thus, the transmission factor or transmissibility of a damped forced elastic system, namely, F_t/F is 1.0 when $\left\{1-\left(\frac{\omega}{\omega_n}\right)^2\right\}^2=1$—i.e. when $\frac{\omega}{\omega_n}=0$ or $\sqrt{2}$ regardless of ζ. The curves of Fig. 4.6(b) show that up to $\omega/\omega_n=\sqrt{2}$ the transmitted force is greater than that for rigid mounting, and beyond $\omega/\omega_n=\sqrt{2}$ the ratio F_t/F is <1.0. Also, beyond $\omega/\omega_n=\sqrt{2}$, an increase in damping increases the ratio F_t/F and is, therefore, detrimental except when

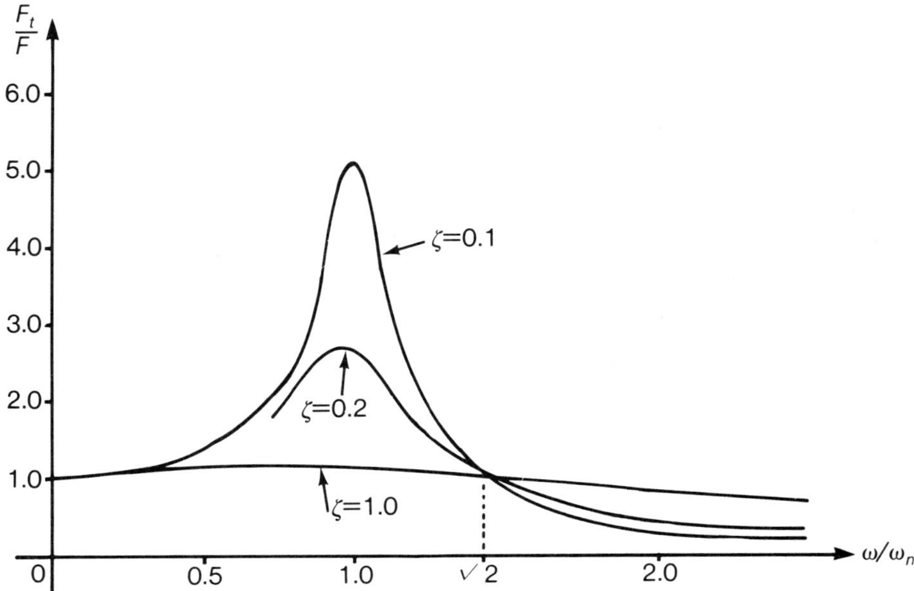

Fig. 4.6(b) Transmission factor or transmissibility curves

$\omega/\omega_n < \sqrt{2}$. For smooth operation, however, $\dfrac{\omega}{\omega_n} > \sqrt{2}$. When $\zeta \to 0$ and $\omega/\omega_n > \sqrt{2}$ the transmissibility $F_t/F \to \dfrac{1}{\left(\dfrac{\omega}{\omega_n}\right)^2 - 1}$ which approaches zero when ω/ω_n is large.

From Fig. 4.6(a) we deduce that the transmitted force, F_t, lags the impressed force, F, by an angle

$$\psi = \beta - \theta = \tan^{-1}\left\{\frac{2\zeta\left(\dfrac{\omega}{\omega_n}\right)^3}{1 - \left(\dfrac{\omega}{\omega_n}\right)^2 + \left(2\zeta\dfrac{\omega}{\omega_n}\right)^2}\right\}$$

It is often necessary to mount rotary machinery (e.g. engines in automobiles) so that there is as little vibration as possible transmitted to the supports. If the forcing frequency, ω, varies, the choice of mountings must be a compromise in that there must be sufficient damping to limit the amplitude and force transmitted while passing through resonance frequency and insufficient to add unacceptably to the force transmitted at the higher operating speeds (see Figs. 4.6(b) and 4.4(f)). Rubber mountings are frequently used as constituting a combination of spring and dashpot—the rubber being of a sufficient size to allow of adequate heat-transfer to surroundings owing to the rise in temperature caused by the energy dissipated in damping. A model for this problem is shown diagrammatically in Figs. 4.1(e) and 4.4(k). It is, however, unusual in actual machinery to find that the entire vibratory motion takes place only in one direction of translation. Dynamic couples are often also present and tend to produce a rocking motion or horizontal as well as vertical vibrations (see Figs. 2.6(d) and 4.1(c)). Thus, the idealized nature of the mathematical model should always be kept in mind when evaluating transmissibility based on rigidity of bodies and massless springs which are fictitious in reality; and, although the analysis of a single degree-of-freedom system is adequate for illustrating the principles involved in vibration isolation, it is not sufficient for appraising many such problems realistically because a rigid body on springs may have six degrees-of-freedom (see Fig. 5.3(1)). In such cases we must resort to the use of computers as labour-saving devices.

EXAMPLE 4.6(i)—illustrative of imbalance and force transmitted to foundations
A machine of mass 100 kg supported on springs of total stiffness 700 N/mm has a rotating imbalanced force of 350 N at a speed of 3000 rev/min. If the damping factor is 0.2, determine (i) the amplitude caused by the imbalance and its phase angle, (ii) the transmissibility, (iii) the actual force transmitted and its phase angle.

Angular speed $\omega = 2\pi \times 50$ rad/s. Undamped natural frequency of system $\omega_n = \sqrt{(k/m)}$ i.e.

$$\omega_n = \sqrt{\left\{700 \frac{\text{N}}{\text{mm}} \times \frac{1}{100\,\text{kg}} \left[\frac{\text{kg}\,10^3\,\text{mm}}{\text{N}\,\text{s}^2}\right]\right\}} = 83.7\,\text{rad/s}$$

Hence, since $\omega/\omega_n = \dfrac{100\,\pi}{83.7} = 3.75$,

(i) amplitude

$$X = \frac{F/k}{\sqrt{\left(\left\{1 - \left(\dfrac{\omega}{\omega_n}\right)^2\right\}^2 + \{2\zeta\omega/\omega_n\}^2\right)}}$$

$$= \frac{350\,\text{N}\,\text{mm}}{700\,\text{N}} \frac{1}{\sqrt{((1-14)^2 + (1.5)^2)}} = \frac{\text{mm}}{26.2}$$

and

$$\beta = \tan^{-1}\left\{\frac{2\zeta(\omega/\omega_n)}{1 - (\omega/\omega_n)^2}\right\} = \tan^{-1}\left(\frac{1.5}{-13}\right) = 173^\circ 24'$$

(ii) transmissibility

$$\frac{F_t}{F} = \sqrt{\left(\frac{1 + (2\zeta\omega/\omega_n)^2}{\{1 - (\omega/\omega_n)^2\}^2 + \{2\zeta\omega/\omega_n\}^2}\right)} = \sqrt{\frac{3.25}{171.25}} = \frac{1}{7.25}$$

(iii) force transmitted

$$F_t = \frac{F}{7.25} = \frac{350}{7.25}\,\text{N} = 48.25\,\text{N}$$

and phase angle

$$\psi = \tan^{-1}\left\{\frac{2\zeta(\omega/\omega_n)^3}{1 - (\omega/\omega_n)^2 + (2\zeta\omega/\omega_n)^2}\right\}$$

$$= \tan^{-1}\left\{\frac{0.4 \times 52.7}{-13 + 2.25}\right\} = 118^\circ 36'.$$

EXAMPLE 4.6(ii)—illustrative of reduction in force transmitted to foundations

A machine having an out-of-balance force of frequency 13 Hz is mounted on rubber pads which have a static deflection of 7.5 mm. Estimate the percentage reduction in transmissibility due to the pads when damping is negligible and $\omega/\omega_n > \sqrt{2}$.

When $\zeta \to 0$ and $\omega/\omega_n > \sqrt{2}$ the transmissibility is $\dfrac{F_t}{F} = \dfrac{1}{(\omega/\omega_n)^2 - 1}$. Also, if d is the static deflection of the pad-mountings, their stiffness $k = \dfrac{mg}{d}$ and the natural angular frequency is $\omega_n = \sqrt{(k/m)} = \sqrt{(g/d)}$. Hence, $F_t/F = \dfrac{1}{\omega^2 \dfrac{d}{g} - 1}$ and the reduction in the maximum force transmitted to the foundations owing to the introduction of the flexible mountings is $F - F_t$.

Thus, the reduction in transmissibility is $\dfrac{F - F_t}{F} = 1 - \left(\dfrac{g}{\omega^2 d - g}\right)$ which, for this case where $\omega = 13\,\text{Hz}\left[\dfrac{2\pi}{\text{Hz s}}\right] = 26\,\pi/\text{s}$ and $d = 7.5\,\text{mm}$, is

$$\frac{F - F_t}{F} = 1 - \frac{9.81\,\text{m/s}^2}{(50 - 9.81)\text{m/s}^2} = 1 - 0.244 = 75.6\%.$$

EXAMPLE 4.6(iii)—illustrative of energy dissipated by damping

Assuming equal energy dissipation per cycle regardless of type of damping in a dynamic system, estimate the equivalent viscous damping in steady-state forced vibration.

In the case of light viscous damping, the amplitude of forced vibration at resonance (see Fig. 4.4(d)) is $X = F/c\omega_n$ and the resonant frequency ω_r approaches $\omega_n = \sqrt{(k/m)}$. No such simple expression exists for other forms of damping (see Sections 3.1 and 3.8) since, e.g. in Coulomb and solid damping, there is difficulty in expressing friction force in the equation of motion over a complete cycle (see Section 3.8). Only in the case of viscous damping, where the friction force is proportional to velocity, does the motion

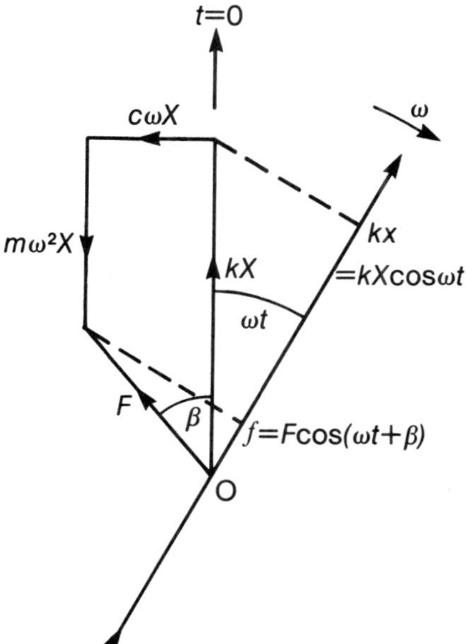

Fig. 4.6(c) Viscous damping phase diagram of forces

proves to be harmonic with sinusoidal excitation. If, however, regardless of the type of damping, we assume that the amplitude at resonance can be represented by $X = \frac{F}{c_e \omega_n}$, where c_e represents the equivalent viscous damping coefficient, we can attempt to find c_e on the basis of equal energy dissipation. In general, for harmonic excitation (referring to Fig. 4.6(c)), the work done per cycle is given by

$$\int_{t=0}^{t=2\pi/\omega} f\frac{dx}{dt}dt = \int_0^{2\pi/\omega} F\cos(\omega t + \beta)X\omega \sin \omega t \, dt$$

i.e. $W = \pi F X \sin \beta$ = energy dissipated/cycle. Hence, when the system is in resonance (i.e. $\beta = 90°$) the work per cycle $= \pi F X = \pi c_e \omega_n X^2$.

This can be equated with the energy dissipated per cycle for any other form of damping determined by assuming harmonic motion and integrating over a cycle (see Example 3.1(i)).

In all damped systems the applied force does work against friction in producing a vibration. The energy dissipated by the transient is negligible since the transient usually vanishes quickly. The dissipation of energy by friction will result in a hot system if heat-transfer to surroundings is inadequate.

EXAMPLE 4.6(iv)—illustrative of the use of an equivalent viscous damping coefficient
Determine the equivalent viscous damping coefficient of a damping force proportional to the square of the velocity assuming harmonic motion $x = X\cos\omega t$. Also find the resonant amplitude.

Symbolically the friction force at any instant may be represented by $f = \pm a\dot{x}^2$, where a is constant and the + sign being used when $\dot{x}$ is negative, and viceversa. Hence, for half a cycle while $\dot{x}$ is positive and doubling we get the energy dissipated/cycle $= 2\int_{-X}^{X} a\dot{x}^2 dx = -2a\omega^2 X^3 \int_{\pi}^{0} \sin^3 \omega t d(\omega t) = \frac{8}{3}a\omega^2 X^3$, which, at resonance, is $\frac{8}{3}a\omega_n^2 X^3 = \pi F X = \pi c_e \omega_n X^2$.

Hence, the equivalent viscous damping coefficient is $c_e = \frac{8}{3\pi}a\omega_n X$ and the resonant amplitude is $X = \sqrt{\left(\frac{3\pi F}{8a\omega_n^2}\right)}$.

For suggested further reading and study, see Appendix D, (6) and (7).

EXERCISES ON CHAPTER 4

1. A cantilever reed has a mass, m, on the end. Derive an equation for the ratio of the amplitude, z, and that of its base which is subjected to harmonic motion $X\sin\omega t$ in the direction perpendicular to the reed.
2. A machine-part vibrates against viscous damping with a resonant amplitude of 12 mm and periodic time 0.2 s caused by a harmonic force of maximum value 24 N. Estimate the damping coefficient.
3. If the machine-part in Exercise 2 has a mass of 2 kg and is freed from the effects of viscous damping, and then excited by a harmonic force of frequency 4 Hz, calculate the increase in amplitude. Also, calculate the elastic coefficient or stiffness of the system.
4. A mass of 5 kg suspended by a spring of stiffness 1 N/mm is forced to oscillate by an

impressed harmonic force of maximum value 10 N against viscous damping of magnitude 0.075 Ns/mm. Find the frequency, amplitude and phase angle (a) for the resonant state, and (b) when peak amplitude occurs.

5. Plot non-dimensionally for damping factors (ζ) 0.01, 0.1, 0.2, 0.5, 1.0 and 2.0, curves of displacement ratio $\left(\dfrac{X}{F/k}\right)$ and phase angle (β) against angular frequency ratio (ω/ω_n) up to 3 for an m, k, c system forced to oscillate in s.h.m. Also, plot a polar diagram of displacement ratio against phase lag, and calculate the values of frequency ratios for maximum displacement ratios.
6. A counter-rotating eccentric-mass exciter produces forced oscillations of a spring-supported mass against a damping force assumed proportional to velocity. At the speed of resonance, an amplitude of 15 mm was recorded, and at a considerably higher speed the amplitude approached the steady value of 2 mm. Estimate the damping factor of the system. (Consult Fig. 4.4(k)).
7. A refrigerator unit of mass 30 kg has an out-of-balance element rotating at 10 rev/s. Calculate the stiffness of springs to support it so that only 10 per cent of the out-of-balance force is transmitted to the foundations.
8. A set of mounted instruments weighing 197.2 N is supported on four rubber blocks subjected to vibration from a rotary engine. If each rubber block has a stiffness of 8 N/mm, calculate the natural frequency of the instrument panel and the percentage of engine-vibration transmitted to the panel when the speed of the engine is 31.8 rev/s and 63.6 rev/s, assuming that the rubber has a damping factor (a) zero and (b) 0.1.
9. A machine of mass 100 kg mounted on springs of total stiffness 1 kN/mm with viscous damping of $\zeta = 0.2$ has a piston of mass 2 kg reciprocating vertically with a stroke of 80 mm at 50 cycle/s. Assuming harmonic motion, determine (i) the amplitude of motion of the machine and its phase angle relative to the exciting force, (ii) the transmissibility ratio, and (iii) the force transmitted to the foundations together with its phase angle relative to the exciting force.

5

Forced vibrations with two degrees of freedom

5.1 UNDAMPED FORCED OSCILLATIONS WITH TWO DEGRESS OF FREEDOM

Following on the cases of undamped natural oscillations with two degrees of freedom dealt with in Section 2.6 we now consider the undamped forced two degree-of-freedom case with particular reference to a simple vibration neutralizer (Fig. 5.1(a)), to be followed in Section 5.2 by a harmonic damper or absorber in which the addition of viscous damping to the auxiliary-mass neutralizer making the analysis more complicated.

We have seen in Section 4.1 that if a spring-mass system with one degree of-freedom has a harmonic force of maximum value F and frequency ω imposed upon it, and that if ω is near the natural frequency ω_n of the system (i.e. approaching resonance) the relatively large amplitude, $\dfrac{F}{m(\omega_n^2 - \omega^2)}$, can be reduced by changing ω_n as a result of altering the mass, m, and/or the spring-rating, k, of the system—thus avoiding the complexity of an added vibration neutralizer. If, however, this is inconvenient or undesirable, a neutralizer (not 'damper' or 'absorber', since no dissipation or absorption of energy takes place) in the form of a spring-supported auxiliary mass or counter-mass may be attached to the main mass in order to create a different natural frequency of the modified system.

Although in practical design problems, a machine element, whose vibration needs to be reduced, will usually have distributed mass, it is expedient in analysis to set up a corresponding 'lumped mass' single degree-of-freedom (m_1, k_1, f) system of Fig. 5.1(a) to which may be added an auxiliary (m_2, k_2) neutralizer system to represent the total mathematical model in order to minimize the complexities in the initial analysis which would, otherwise need the aid of a computer.

In its simplest form the neutralizer consists of a mass and spring combination of the same frequency (ω_{n2}) as that (ω) of the disturbing force or as that (ω_{n1}) of the original one degree-of-freedom system. The addition of the auxiliary mass changes the original system into one with two degrees-of-freedom which, necessarily, has two natural frequencies—one above and one below that of the original system. With a correct design of neutralizer (i.e. when $\omega_{n2} = \omega$) the amplitude of the original system can be made zero in accordance with the following analysis. A similar design can be carried out for undamped torsional systems. In both linear and torsional systems, vibration neutralizers are designed merely to induce a force or torque to oppose the exerting force or torque, respectively, and so diminish unwanted vibration. We shall see, however, that such a harmonic neutralizer is not practicable in systems with a wide range of frequency because it can only be designed to suppress resonance at one frequency whilst allowing resonance at two other frequencies (see Fig. 5.1(c)).

Consider the total system (with two degrees of freedom) shown diagrammatically in Fig. 5.1(a). The main mass m_1 subjected to a harmonic force $f = F \cos \omega t$ will, after subsidence of any initial vibration, induce the same frequency, ω, in the auxiliary mass m_2. Thus, if the response is assumed to

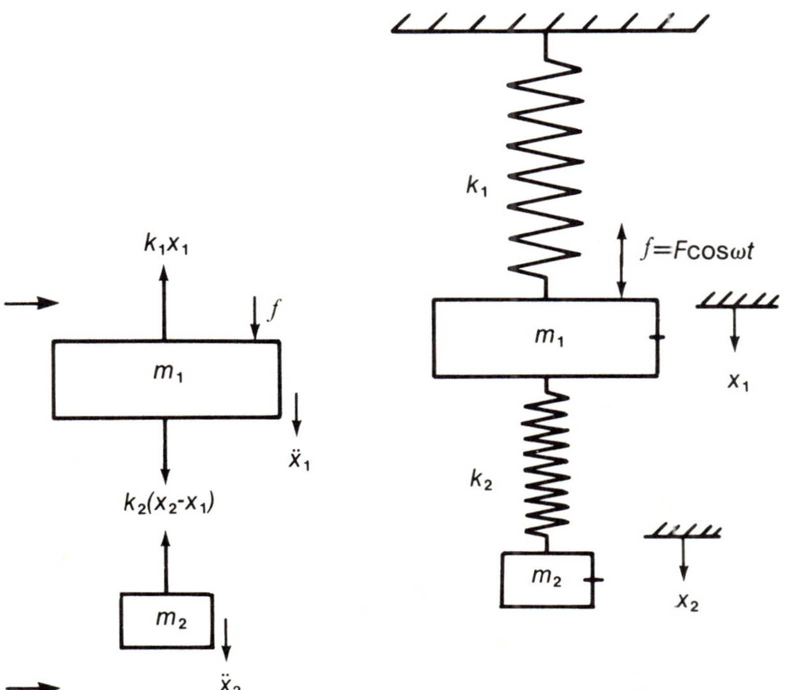

Fig. 5.1(a) Auxiliary mass vibration neutralizer

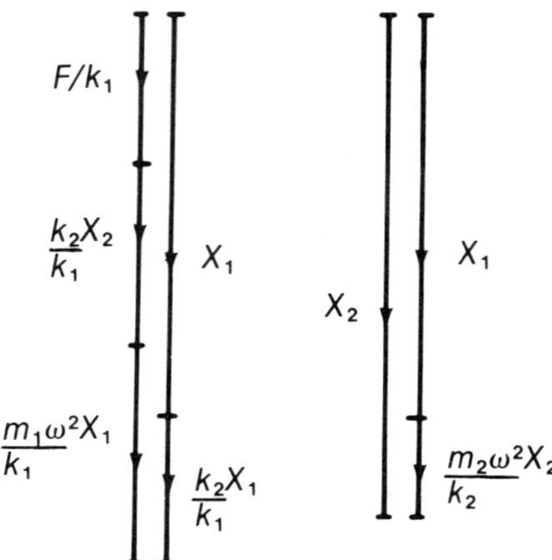

Fig. 5.1(b) Diagrammatic of equations (1a) and (2a)

be $x_2 = X_2 \cos(\omega t + \beta)$, the vector diagram of forces won't close unless $\beta = 0$ (see Fig. 5.1(b)). Hence, the responses to the applied harmonic force are simple harmonic motions $x_1 = X_1 \cos \omega t$ and $x_2 = X_2 \cos \omega t$, where X_1 is the amplitude of the main system, and required to be minimized or eliminated. The equations of motion for the two masses are:

$$m_1 \ddot{x}_1 = f - k_1 x_1 + k_2(x_2 - x_1) \tag{1}$$

and

$$m_2 \ddot{x}_2 = -k_2(x_2 - x_1) \tag{2}$$

Alternatively, as shown on Fig. 5.1(b):

$$-\frac{m_1}{k_1} X_1 \omega^2 + \left(1 + \frac{k_2}{k_1}\right) X_1 = \frac{k_2}{k_1} X_2 + \frac{F}{k_1} \tag{1a}$$

and

$$-\frac{m_2}{k_2} X_2 \omega^2 + X_2 = X_1 \tag{2a}$$

which can be written as:

$$X_1 \left\{ 1 + \frac{k_2}{k_1} - \left(\frac{\omega}{\omega_{n1}}\right)^2 \right\} - \frac{k_2}{k_1} X_2 = \frac{F}{k_1} = X_{st1} \tag{1b}$$

and

$$X_2\left\{1-\left(\frac{\omega}{\omega_{n2}}\right)^2\right\}=X_1 \tag{2b}$$

where X_{st1} is the static deflection (or zero frequency deflection) of m_1 caused by a steady force F (not to be confused with static deflection-caused by the weight of mass m_1), and $\omega_{n1}^2=k_1/m_1$ and $\omega_{n2}^2=k_2/m_2$—i.e. ω_{n1} and ω_{n2} would be the natural frequencies of the two systems when disconnected but which cannot exist in a real sense when the systems are connected together. In matrix form (see Appendix B3), equations (1a) and (2a) may be written

$$\begin{bmatrix}(k_1+k_2-m_1\omega^2) & -k_2\\ -k_2 & (k_2-m_2\omega^2)\end{bmatrix}\begin{Bmatrix}X_1\\ X_2\end{Bmatrix}=\begin{bmatrix}F\\ 0\end{bmatrix}.$$

Eliminating X_2 from equations (1b) and (2b) yields

$$\frac{X_1}{X_{st1}}=\frac{1-(\omega/\omega_{n2})^2}{\left\{1+\frac{k_2}{k_1}-\left(\frac{\omega}{\omega_{n1}}\right)^2\right\}\left\{1-\left(\frac{\omega}{\omega_{n2}}\right)^2\right\}-\left(\frac{k_2}{k_1}\right)} \tag{1c}$$

and eliminating X_1 yields

$$\frac{X_2}{X_{st1}}=\frac{1}{\left\{1+\frac{k_2}{k_1}-\left(\frac{\omega}{\omega_{n1}}\right)^2\right\}\left\{1-\left(\frac{\omega}{\omega_{n2}}\right)^2\right\}-\left(\frac{k_2}{k_1}\right)} \tag{2c}$$

Thus, from equations (1c) and (2b), we see that $X_1=0$ when $\omega=\omega_{n2}$, i.e. when the exciting frequency, ω, coincides with the natural frequency of the harmonic auxiliary system 2 the motion of the main mass, m_1, does not simply diminish but ceases altogether. Also, when $\omega=\omega_{n2}$ we deduce from equations (2c) and (1b) that $m_2\omega^2X_2=k_2X_2=-k_1X_{st1}=-F$, i.e. the force k_2X_2 exerted by the spring is equal and opposite to the applied force F—this being the basis of the vibration neutralizer.

The vector diagrams of Fig. 5.1(b) also show that if X_1 is to be zero, then $m_2\omega^2X_2=k_2X_2=-F$, i.e. $\omega^2=k_2/m_2=\omega_n^2$ and, since $m_2=\dfrac{F_2}{\omega^2X_2}$, the size of the auxiliary mass, m_2, will depend on the size of the disturbing force, F, and the maximum force, F, and the maximum amplitude, X_2, allowable.

The ratio $\dfrac{X_1}{X_{st1}}=\dfrac{X_2}{F/k_1}$ represents a magnification factor—being the ratio of actual amplitude to that which would occur if the maximum value of the impressed force were applied as a static load. This ratio (equation 1c) is plotted

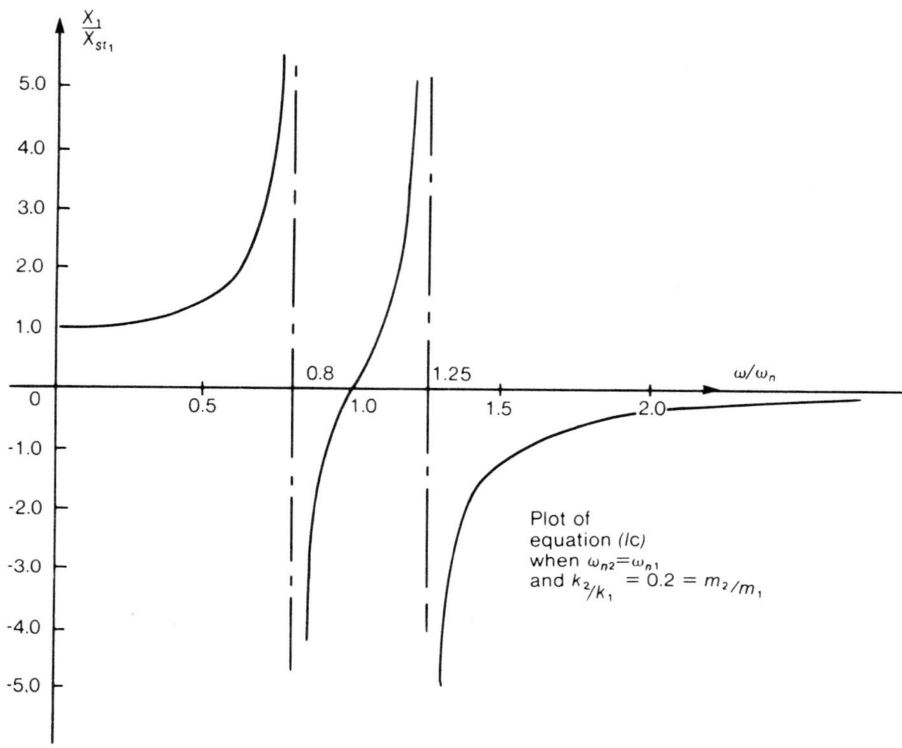

Fig. 5.1(c) Magnification factor against frequency ratio

against ω/ω_{n2} on Fig. 5.1(c) for a particular case, as is $\dfrac{X_2}{X_{st1}} = \dfrac{X_2}{F/k_1}$ (equation 2c) on Fig. 5.1(d).

Resonant frequencies ($X_1 \to \infty$) will occur when the denominator of equation (1c) is zero—i.e. when $\{1 + ra^2 - b^2\}\{1 - (b/a)^2\} - ra^2 = 0$, where $a = \omega_{n2}/\omega_{n1}$; $b = \omega/\omega_{n1}$; $r = m_2/m_1$; and $\dfrac{k_2}{k_1} = \dfrac{m_2}{m_1}.\dfrac{k_2}{k_1}.\dfrac{m_1}{k_1} = r\left(\dfrac{\omega_{n2}}{\omega_{n1}}\right)^2 = ra^2$. Hence $\left(\dfrac{b}{a}\right)^4 - \left\{r + 1 + \dfrac{1}{a^2}\right\}\left(\dfrac{b}{a}\right)^2 + \dfrac{1}{a^2} = 0$, which is a quadratic in $\left(\dfrac{b}{a}\right)^2$ or $\left(\dfrac{\omega}{\omega_{n2}}\right)^2$ and has two roots, namely, $\left(\dfrac{\omega_1}{\omega_{n2}}\right)^2$ and $\left(\dfrac{\omega_2}{\omega_{n2}}\right)^2$ given by $2\left(\dfrac{b}{a}\right)^2 = \left(r + 1 + \dfrac{1}{a^2}\right) \pm \sqrt{\left\{\left(r + 1 + \dfrac{1}{a^2}\right)^2 - \left(\dfrac{2}{a}\right)^2\right\}}$ (see Appendix B2).

The product of the roots is $\left(\dfrac{\omega_1}{\omega_{n2}}\right)^2 \times \left(\dfrac{\omega_2}{\omega_{n2}}\right)^2 = \dfrac{1}{a^2}$, and their sum

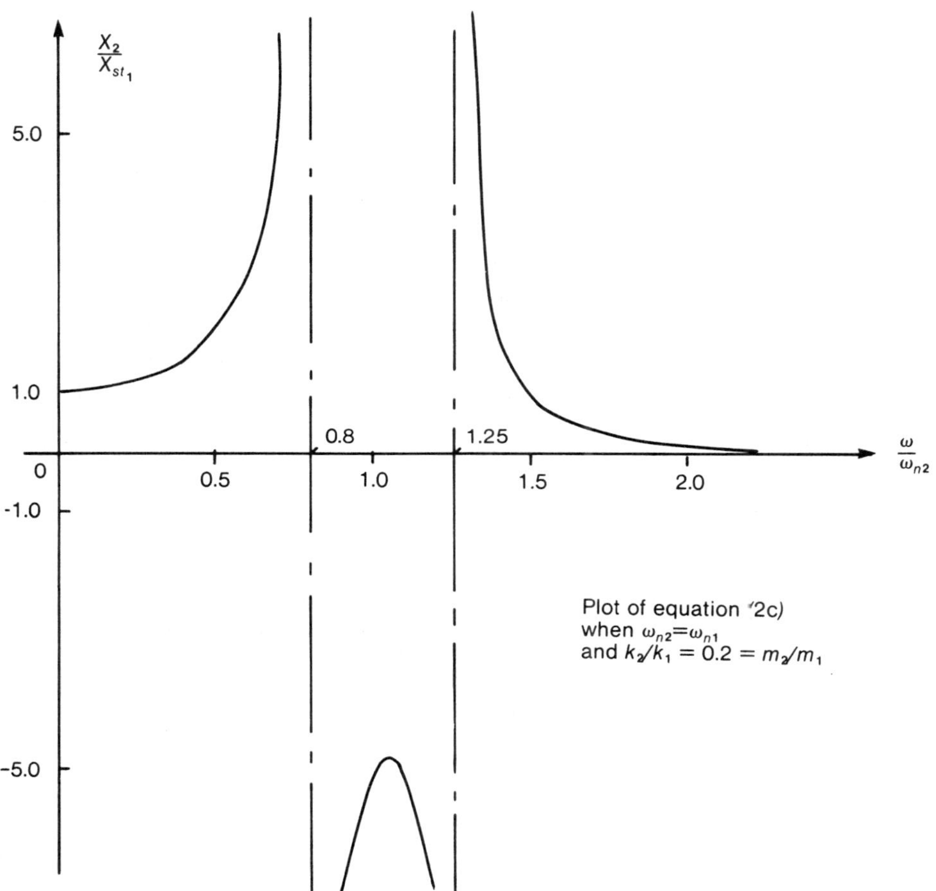

Fig. 5.1(d) Magnification ratio against frequency ratio

$\left(r+1+\dfrac{1}{a^2}\right)$. Thus, when $\left(\dfrac{\omega}{\omega_{n2}}\right)^4-\left\{\left(\dfrac{\omega_{n1}}{\omega_{n2}}\right)^2+\dfrac{m_2}{m_1}+1\right\}\left(\dfrac{\omega}{\omega_{n2}}\right)^2+\left(\dfrac{\omega_{n1}}{\omega_{n2}}\right)^2=0$ for any values of $\dfrac{\omega_{n2}}{\omega_{n1}}=a$ and $\dfrac{m_2}{m_1}=r$, two values of $\left(\dfrac{\omega}{\omega_{n2}}\right)^2$ can be found as the resonant frequency ratios.

Alternatively, they can be estimated from the graph of equation (1c) of which Fig. 5.1(c) is a particular case, namely, the auxiliary system tuned to the natural frequency of the main system $\omega_{n2}=\omega_{n1}$, i.e. $a=1$ and $\dfrac{m_2}{m_1}=0.2=r$ which gives resonance when $\left(\dfrac{b}{a}\right)^4-2.2\left(\dfrac{b}{a}\right)^2+1=0$ or when $\left(\dfrac{b}{a}\right)^2=$

$\left(\frac{\omega}{\omega_{n2}}\right)^2 = 1.1 \pm \sqrt{(1.21-1)} = 1.56$ and 0.64. Hence, when $\frac{\omega}{\omega_{n2}} = 0.8$ and 1.25, $\frac{X_1}{X_{st\,1}}$ approaches infinity as the graph shows, and from which it will also be seen that this particular vibration neutralizer is completely effective when the frequency (ω) of the applied harmonic force is equal to the natural frequency (ω_{n2}) of the absorber system.

Since the dynamic vibration neutralizer is tuned to one particular frequency ($\omega_{n2} = \omega$), it is only effective over a narrow band of frequencies and can be completely effective (i.e. $X_1 = 0$) when applied to synchronous machines and reciprocating tools of constant operating frequency (ω). A satisfactory operating range is likely to be where $X_1/X_{st\,1}$ is less than 1 as indicated in Fig. 5.1(c). Also, we may note that when $\frac{\omega}{\omega_{n2}} = 1$, $\frac{X_1}{X_{st\,1}} = 0$ but $\frac{X_2}{X_{st\,1}} = 5$ as indicated on Fig. 5.1(d).

EXAMPLE 5.1(i)—illustrative of the size of a vibration neutralizer

An instrument panel is caused to vibrate by an out-of-balance machine. The two frequencies at which severe vibration of the panel occurred were 20 Hz and 25 Hz. Estimate the mass ratio of a simple spring-mass neutralizer if attached to the panel and tuned so that $\omega_{n2} = \omega_{n1}$.

Referring to the analysis of Fig. 5.1(a),

$$\left(\frac{\omega_1}{\omega_{n2}}\right)^2\left(\frac{\omega_2}{\omega_{n2}}\right)^2 = \frac{1}{a^2} \quad \text{and} \quad \left(\frac{\omega_1}{\omega_{n2}}\right)^2 + \left(\frac{\omega_2}{\omega_{n2}}\right)^2 = \left(r + 1 + \frac{1}{a^2}\right),$$

where $a = \frac{\omega_{n2}}{\omega_{n1}} = 1$ in this case, $r = \frac{m_2}{m_1}$, and ω_1 and ω_2 are resonant frequencies.

For this panel, $\omega_1 = 20$ Hz and $\omega_2 = 25$ Hz Hence, $\frac{20 \times 25}{\omega_{n2}^2} = 1$, i.e. $\omega_{n2} = 22.4$ Hz and

$$(r + 1 + 1) = \left(\frac{20}{22.4}\right)^2 + \left(\frac{25}{22.4}\right)^2 = \frac{400 + 625}{500} = 2.05$$

Thus,

$$r = \frac{m_2}{m_1} = 0.05$$

EXAMPLE 5.1(ii)—illustrative of amplitude ratio in a system with two degress of freedom

Determine the natural frequencies of the combined system shown in Fig. 5.1(a) when $m_1 = 1.75$ kg, $k_1 = 3.5$ N/mm and $m_2 = 0.875$ kg, $k_2 = 1.75$ N/mm. Also, determine the ratio of the amplitudes of the two masses at the higher and lower natural frequencies.

Referring to Fig. 5.1(a) without the impressed harmonic force, the equations of motion are: $m_1\ddot{x}_1 = -k_1x_1 + k_2(x_2 - x_1)$ and $m_2x_2 = -k_2(x_2 - x_1)$. For principal modes of oscillation, $x_1 = X_1 \cos \omega t$ and $x_2 = X_2 \cos \omega t$. Hence,

$$X_1\left\{1 + \frac{k_2}{k_1} - \left(\frac{\omega}{\omega_{n1}}\right)^2\right\} = \frac{k_2}{k_1}X_2,$$

and

$$X_2\left\{1 - \left(\frac{\omega}{\omega_{n2}}\right)^2\right\} = X_1, \quad \text{since } \omega_{n1}^2 = \frac{k_1}{m_1} \quad \text{and} \quad \omega_{n2}^2 = \frac{k_2}{m_2}.$$

Therefore,

$$\frac{X_1}{X_2} = \frac{k_2/k_1}{1 - k_2/k_1 - \left(\dfrac{\omega}{\omega_{n1}}\right)^2} = 1 - \left(\frac{\omega}{\omega_{n2}}\right)^2,$$

and

$$r\left(\frac{\omega_{n2}}{\omega_{n1}}\right)^2 = \left\{1 - \left(\frac{\omega}{\omega_{n2}}\right)^2\right\}\left\{1 - r\left(\frac{\omega_{n2}}{\omega_{n1}}\right)^2 - \left(\frac{\omega}{\omega_{n1}}\right)^2\right\},$$

where

$$r = \frac{m_2}{m_1} \quad \text{and} \quad \frac{k_2}{k_1} = r\left(\frac{\omega_{n2}}{\omega_{n1}}\right)^2.$$

In the particular case, $r = 1/2$ and $\dfrac{k_2}{k_1} = \dfrac{1}{2}$; hence, $\left(\dfrac{\omega_{n2}}{\omega_{n1}}\right) = 1$, and

$$\omega_{n1}^2 = \frac{k_1}{m_1} = 3.5\,\frac{\text{N}}{\text{mm}} \times \frac{1}{1.75\,\text{kg}}\left[\frac{\text{kg}\,10^3\,\text{mm}}{\text{N}\,\text{s}^2}\right] = \frac{2000}{\text{s}^2} = \omega_{n2}.$$

Therefore,

$$\tfrac{1}{2} \times 1 = \left\{1 - \frac{\omega^2}{2000/\text{s}^2}\right\}\left\{1 - \frac{1}{2} - \frac{\omega^2}{2000/\text{s}^2}\right\}$$

or $(\omega s)^4 - 5 \times 10^2(\omega s)^2 + 4 \times 10^6 = 0$, i.e. $(\omega s)^2 = 4 \times 10^3$ or 1×10^3 (see Appendix B2). Hence, $\omega = 63.3/\text{s}$ or $31.6/\text{s}$ and $f = \dfrac{\omega}{2\pi} = 10$ Hz or 5 Hz for the two principal modes of vibration of this system with two degrees of freedom.

The ratio of the amplitudes is $\dfrac{X_1}{X_2} = 1 - \left(\dfrac{\omega}{\omega_{n2}}\right)^2 = 1 - \tfrac{4}{2} = -1$ for the higher mode, and $1 - \tfrac{1}{2} = \tfrac{1}{2}$ for the lower mode.

5.2 DAMPED FORCED OSCILLATIONS WITH TWO DEGREES OF FREEDOM

Although every dynamic system has a certain amount of damping, it is often necessary to add damped spring-supported mass devices so tuned as to reduce or eliminate vibrations, e.g. resonant oscillations of beams can be reduced to insignificance by such a device attached at mid-span; and the severe torsional vibration that would exist in the crankshaft of an engine having a natural frequency within the range of the operating frequency of the engine, can be reduced by coupling a fly wheel to the shaft through a spring and damper. However, the adequacy of damping or otherwise in any design can only be proved by tests on the particular machine or structure; and this applies to all theoretical predictions.

Consider the general case shown diagrammatically in Fig. 5.2(a) vibrating horizontally, and in which we assume that the masses m_1 and m_2 are concentrated, that springs are massless and that damping all takes place in the dashpot, and is viscous. If the angular frequency of the forced vibration is ω, we may use ratios: $m_2/m_1 = r$; $\omega_{n2}/\omega_{n1} = a$; $\omega/\omega_{n1} = b =$ frequency of forced vibration/natural frequency of main mass, and $\dfrac{\omega}{\omega_{n2}} = \dfrac{\omega}{\omega_{n1}} \cdot \dfrac{\omega_{n1}}{\omega_{n2}} = \dfrac{b}{a}$ (as in Section 5.1). Also, let $\zeta = \dfrac{c}{2\sqrt{(m_2 k_2)}} = \dfrac{c\omega_{n2}}{2k_2}$, where $\omega_{n1}^2 = k_1/m_1$ and $\omega_{n2}^2 = k_2/m_2$ then, assuming harmonic motion $x_1 = X_1 \cos(\omega t + \beta)$ and $x_2 = X_2 \cos \omega t$ in response to the harmonic force f, the equations of motion for the two masses are:

$$f - k_1 x_1 - k_2(x_1 - x_2) - c(\dot{x}_1 - \dot{x}_2) = m_1 \ddot{x}_1 \tag{1}$$

and

$$k_2(x_1 - x_2) + c(\dot{x}_1 - \dot{x}_2) = m_2 \ddot{x}_2 \tag{2}$$

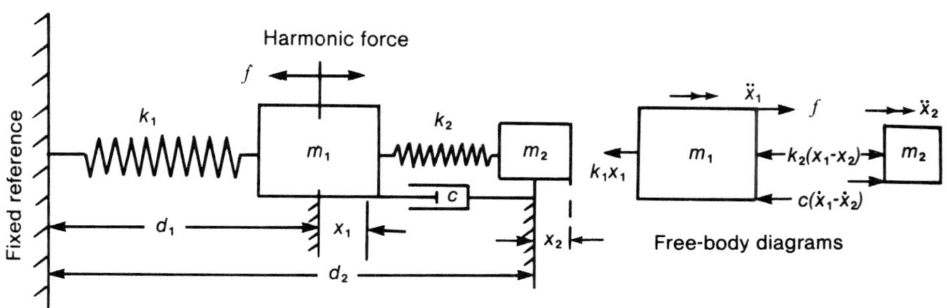

Fig. 5.2(a) Vibration damper

Hence, using equation (2) in the form

$$m_2\ddot{x}_2 + c\dot{x}_2 + k_2x_2 = c\dot{x}_1 + k_2x_1 \tag{2a}$$

and equivalents $\dfrac{c\omega}{k_2} = 2\zeta\dfrac{\omega}{\omega_{n2}}$; $\dfrac{m_2}{k_2}\omega^2 = \left(\dfrac{\omega}{\omega_{n2}}\right)^2$; $\dfrac{c\omega X_1}{k_2X_2} = 2\zeta\left(\dfrac{\omega}{\omega_{n2}}\right)\dfrac{X_1}{X_2}$, the phase diagrams for mass m_2 are shown in Fig. 5.2(b).

From the geometry of the non-dimensional (Fig. 5.2(b)) diagram, we see that

$$z^2 = \{1-(\omega/\omega_{n2})^2\}^2 + \{2\zeta(\omega/\omega_{n2}\}^2 = \left(\frac{X_1}{X_2}\right)^2 + \left(2\zeta\left(\frac{\omega}{\omega_n}\right)\frac{X_1}{X_2}\right)^2$$

Alternatively,

$$z^2 = \{1-(b/a)^2\}^2 + \{2\zeta b/a\}^2 = \left(\frac{X_1}{X_2}\right)^2\{1+2\zeta b/a\}^2$$

Also, $\cos\beta = \cos(\psi-\alpha) = \cos\psi\cos\alpha + \sin\psi\sin\alpha$

i.e.

$$\cos\beta = \frac{X_1}{z^2X_2}\left\{\left(1-\left(\frac{\omega}{\omega_{n2}}\right)^2\right)+\left(2\zeta\frac{\omega}{\omega_{n2}}\right)^2\right\}$$

or

$$\frac{X_2}{X_1}\cos\beta = \frac{1}{z^2}\{(1-(b/a)^2)+(2\zeta b/a)^2\}$$

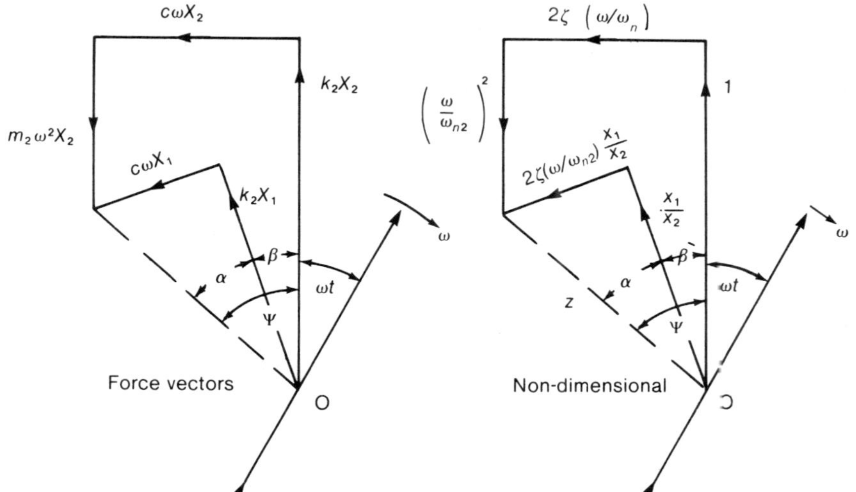

Fig. 5.2(b) Vector phase diagrams of equation (2a)

Considering a re-arranged equation (1), namely,

$$m_1\ddot{x}_1 + c\dot{x}_1 + (k_1 + k_2)x_1 = f + c\dot{x}_2 + k_2x_2 \tag{1a}$$

we can draw the vector diagram Fig. 5.2(c). Also, by adding equations (1a) and (2a) we get

$$m_2\ddot{x}_2 + m_1\ddot{x}_1 + k_1x_1 = f \tag{3}$$

which has a common term $(m_2\ddot{x}_2)$ with equation (2a), and a common vector with $m_2\omega^2X_2$ on the force diagram of Fig. 5.2(b). Using this common vector, $m_2\omega^2X_2$, we may superimpose the vector diagram of forces of equation (3) on that of Fig. 5.2(b) to get Fig. 5.2(d) which yields the maximum value, F, of the impressed harmonic force f. The two components of F, from the geometry of Fig. 5.2(d), are:

$$F_V = m_2\omega^2X_2 - (k_1X_1 - m_1\omega^2X_1)\cos\beta$$

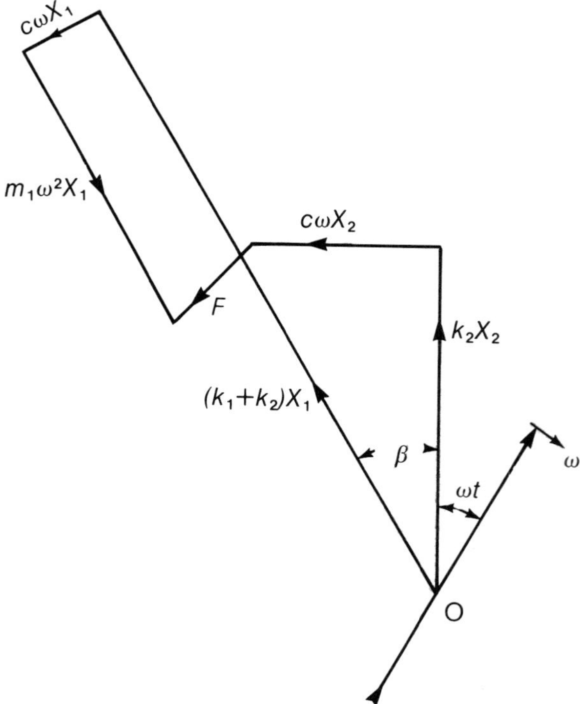

Fig. 5.2(c) Max forces vector diagram for equation (1a)

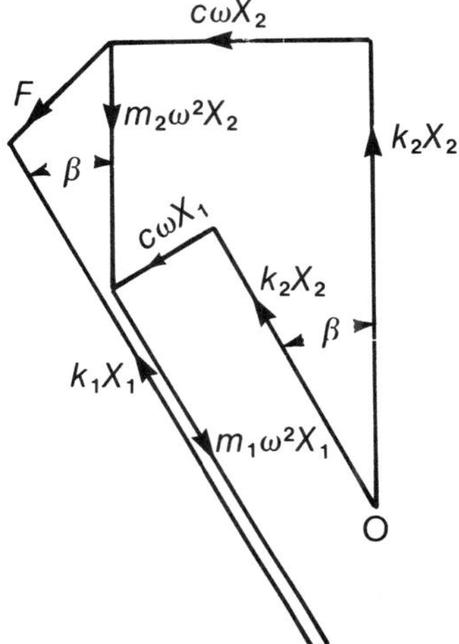

Fig. 5.2(d) Max forces vector diagram for equations (1a), (2a) & (3) superimposed

or

$$\frac{F_V}{k_1 X_1} = \frac{m_2}{m_1}\frac{m_1}{k_1}\omega^2\frac{X_2}{X_1} - \left(1 - \frac{m_1}{k_1}\omega^2\right)\cos\beta$$

$$= rb^2\frac{X_2}{X_1} - (1 - b^2)\cos\beta$$

and

$$F_H = (k_1 X_1 - m_1\omega^2 X_1)\sin\beta$$

or

$$\frac{F_H}{k_1 X_1} = \left(1 - \frac{m_1}{k_1}\omega^2 X_1\right)\sin\beta = (1 - b^2)\sin\beta$$

Hence,

$$\left(\frac{F}{k_1 X_1}\right)^2 = \left\{rb^2\frac{X_2}{X_1} - (1 - b^2)\cos\beta\right\}^2 + \{(1 - b^2)\sin\beta\}^2$$

$$= (rb^2)^2\left(\frac{X_2}{X_1}\right)^2 - 2rb^2\frac{X_2}{X_1}(1 - b^2)\cos\beta + (1 - b^2)^2$$

which, on substituting for $\left(\frac{X_1}{X_2}\right)^2$ and $\left(\frac{X_2}{X_1}\right)\cos\beta$ and z^2, becomes:

$$\left(\frac{zF}{k_1X_1}\right)^2 = (rb^2)^2\left\{1+\left(2\zeta\frac{b}{a}\right)^2\right\} - 2rb^2(1-b^2)\left\{\left(1-\left(\frac{b}{a}\right)^2\right)\right.$$

$$\left.+\left(2\zeta\frac{b}{a}\right)^2\right\} + (1-b^2)^2\left\{\left(1-\left(\frac{b}{a}\right)^2\right)^2+\left(2\zeta\frac{b}{a}\right)^2\right\}$$

$$= \left(2\zeta\frac{b}{a}\right)^2\{rb^2-(1-b^2)\}^2$$

$$+\left\{rb^2-(1-b^2)\left(1-\left(\frac{b}{a}\right)^2\right)\right\}^2$$

Hence,

$$\left(\frac{X_1}{F/k_1}\right)^2 = \frac{(a^2-b^2)^2+(2\zeta ab)^2}{(2\zeta ab)^2\{b^2(1+r)-1\}^2+\{ra^2b^2-(1-b^2)(a^2-b^2)\}^2}$$

from which we can plot graphs of $\dfrac{X_1}{F/k_1}$ against $b=\omega/\omega_{n1}$ for a given $a=\omega_{n2}/\omega_{n1}$ and $r=m_2/m_1$ for a series of values of ζ.

Also, writing $\dfrac{X_1}{F/k_1}=\sqrt{\left\{\dfrac{(2\zeta ab)^2+A}{(2\zeta ab)^2C+D}\right\}}$, then if $A=D/C$ we have $\dfrac{X_1}{F/k_1}=\sqrt{\dfrac{1}{C}}=\pm\left(\dfrac{1}{b^2(1+r)-1}\right)$, which condition makes $\dfrac{X_1}{F/k_1}$ independent of ζ. With this $(AC=D)$ condition, i.e. $(a^2-b^2)\{b^2(1+r)-1\} = \pm\{ra^2b^2-(1-b^2)(a^2-b^2)\}$, the plus sign yields $-b^4r=0$, i.e. $b=\omega/\omega_{n1}=0$ or $\omega=0$; and the minus sign yields $(2+r)b^4-2b^2(ra^2+a^2+1)+2a^2=0$ which (see Appendix B2) yields

$$b^2=\left(\frac{ra^2+a^2+1}{2+r}\right)\pm\sqrt{\left\{\left(\frac{ra^2+a^2+1}{2+r}\right)^2-\left(\frac{2a}{2+r}\right)\right\}}$$

from which (since b cannot be negative, and r and a are always positive) two values of $b=\omega/\omega_{n1}$ can be found which are independent of ζ, i.e. the series of curves for different values of ζ on the graphs of $X/F/k_1$ against ω/ω_{n1} will all intersect at these two values of b.

EXAMPLE 5.2(i)—illustrative of the relative movement of masses in a rolling ship
Two concentrated masses in close proximity in a ship are assumed equivalent to masses m_1 and m_2 connected by a spring k_2, the mass m, being connected to a very massive structure by a second spring k_1—the two springs being in line. There is viscous damping between m_1 and the heavy support, but no damping between m_1 and m_2. Deduce an expression for the relative movement of the masses when an external force $f=F\cos(\omega t+\beta)$ acts on m_1 in the line of the springs.

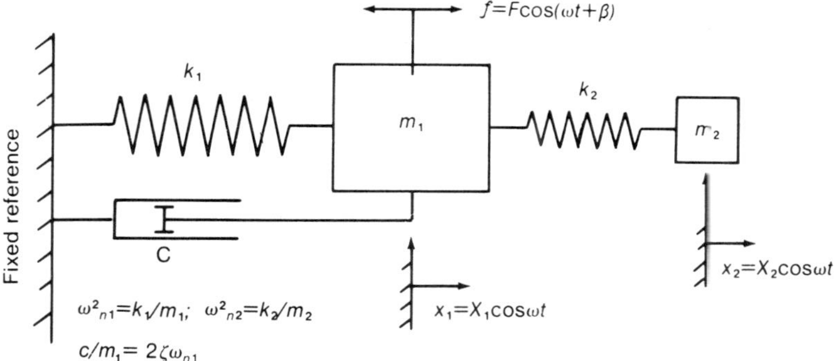

Fig. 5.2(e) Equivalent vibrating system

Referring to Fig. 5.2(e) the equations of motion are:

$$m_2\ddot{x}_2 = -k_2(x_2 - x_1) \text{ or } m_2\ddot{x}_2 + k_2x_2 = k_2x_1 \tag{1}$$

and $m_1\ddot{x}_1 = k_2(x_2 - x_1) - k_1x_1 - c\dot{x}_1 + f$

or

$$m_1\ddot{x}_1 + c\dot{x}_1 + (k_1 + k_2)x = k_2x_2 + f \tag{2}$$

Adding equations (1) and (2) gives:

$$m_2\ddot{x}_2 + m_1\ddot{x}_1 + c\dot{x}_1 + k_1x_1 = f \tag{3}$$

Hence, assuming harmonic motion for x_1 and x_2 in response to f, the phase diagrams of maximum values are as shown in Fig. 5.2(f).

From the geometry of the combined phase or vector diagram for equation (3) we deduce that

$$F^2 = \{k_1X_1 - m_2\omega^2X_2 - m_1\omega^2X_1\}^2 + (c\omega X_1)^2.$$

Also, from the diagram of equation (1) we see that

$$k_2X_2 - m_2\omega^2X_2 = k_2X_1, \text{ i.e. } \frac{X_1}{X_2} = 1 - \left(\frac{\omega}{\omega_{n2}}\right)^2$$

Hence,

$$\left(\frac{F}{k_1X_1}\right)^2 = \left\{1 - \frac{m_2}{m_1}\left(\frac{\omega}{\omega_{n1}}\right)^2\left(\frac{1}{1 - \left(\frac{\omega}{\omega_{n2}}\right)^2}\right) - \left(\frac{\omega}{\omega_{n2}}\right)^2\right\}^2 + \left(2\zeta\frac{\omega}{\omega_{n1}}\right)^2$$

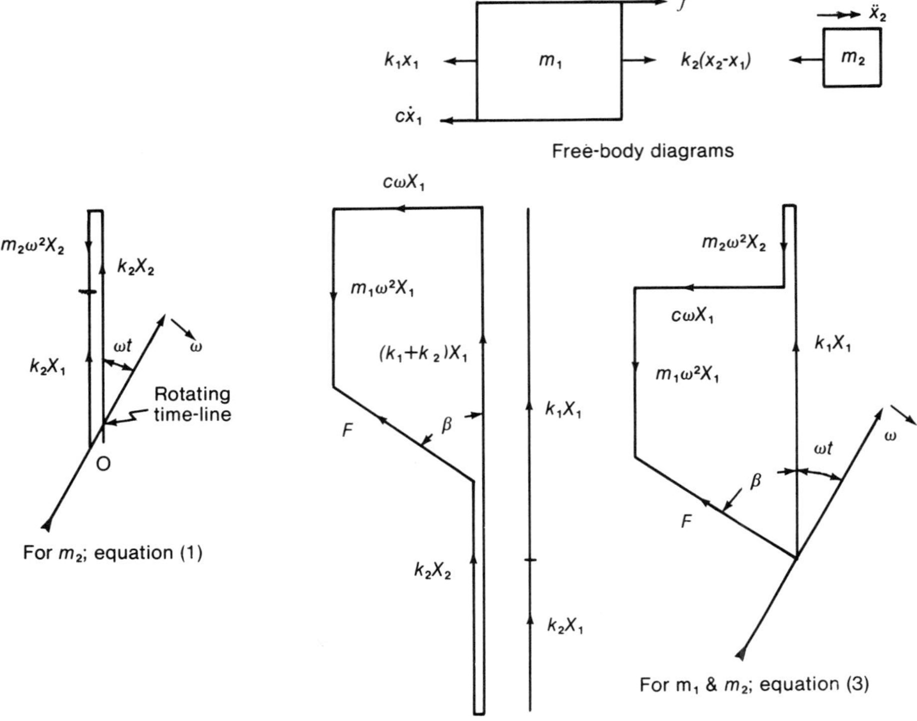

Fig. 5.2(f) Vector diagrams of maximum forces and phases

from which $\dfrac{X_1}{F/k_1}$ and X_1 can be found. Also, $\dfrac{X_2 - X_1}{X_1} = \dfrac{(\omega/\omega_{n2})^2}{1-(\omega/\omega_{n2})^2}$ and the phase angle β is obtained from $\tan\beta = \dfrac{c\omega X_1}{k_1X_1 - m_2\omega^2X_2 - m_1\omega^2X_1}$.

5.3 DAMPED FORCED OSCILLATIONS OF SYSTEMS WITH MOVING SUPPORTS

Three general cases (A, B, C) will be considered in which a mass is forced (or induced) to oscillate because of a vibrating support or a displacement input.

(A) Consider the case of a system which is virtually that of a mass suspended on a spring with viscous damping provided by a stationary guide or dashpot as in Fig. 5.3(a). Assume the support moves in vertical linear harmonic motion and that the mass responds in the fixed dashpot or lubricated guides with simple harmonic motion of the same frequency ω.

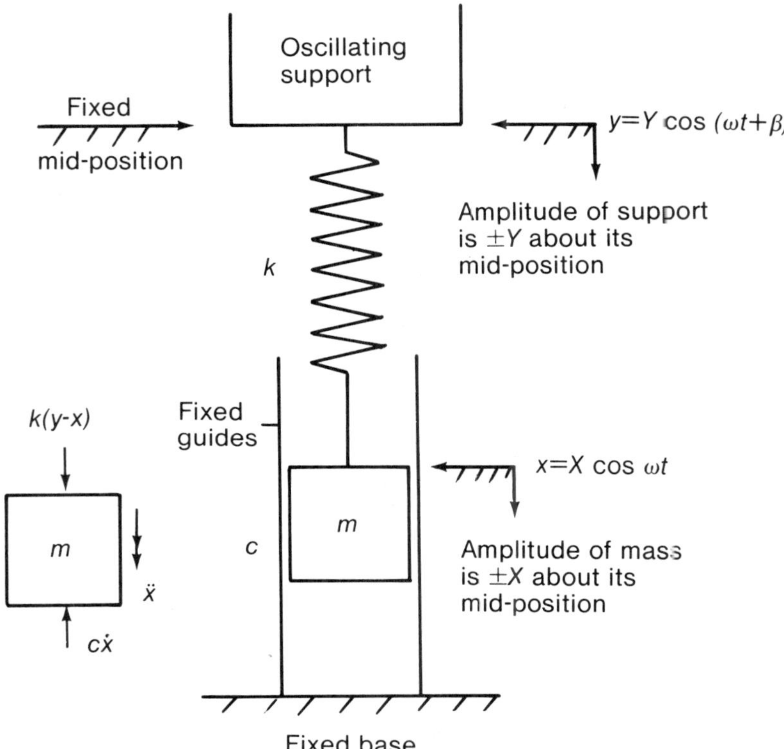

Fig. 5.3(a) Damped forced oscillations caused by a moving support

The ratio of the amplitudes, X/Y, can be deduced by applying the principle of "cause = effect" or force = mass × acceleration which, in symbols, is $k(y-x)-c\dot{x}=m\ddot{x}$ after the natural frequency has disappeared as a result of damping, i.e. $m\ddot{x}+c\dot{x}+kx=ky$ which, on substituting, becomes $m\omega^2 X\cos(\omega t+\pi)+c\omega X\cos(\omega t+\pi/2)+kX\cos\omega t=kY\cos(\omega t+\beta)$. This latter equations can be represented by the vector diagrams shown in Fig. 5.3(b) from which we deduce that

$$Y/X=\surd(\{1-(\omega/\omega_n)^2\}^2+\{2\zeta\omega/\omega_n\}^2)$$

and

$$\tan\beta=\frac{2\zeta\omega/\omega_n}{1-(\omega/\omega_n)^2}$$

where X and Y are measured relative to a fixed foundation on the earth. The curves of displacement ratio and angle of lag for a series of damping factors are the same as those shown on Figs. 4.4(f and g).

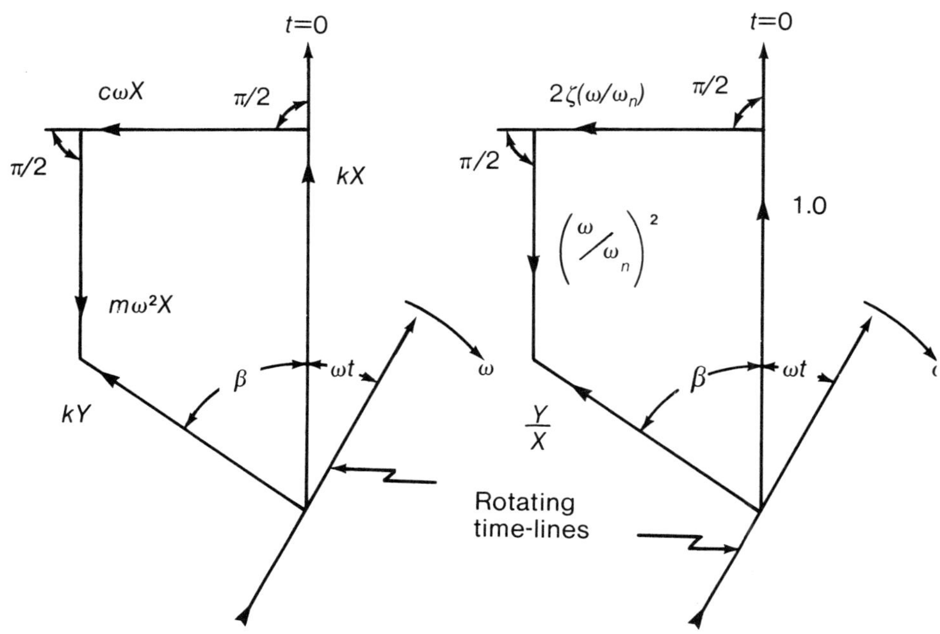

Fig. 5.3(b) Vector diagrams of forces and force ratios

EXAMPLE 5.3(i)—illustrative of oscillations induced by a vibrating support
A mass suspended by a vertical helical spring has a natural frequency of 10 Hz and its motion is damped by viscous damping so that the amplitude of the oscillations is halved in 2 s. If the upper end of the spring is given a vertical s.h.m. of frequency 10 Hz and amplitude Y, estimate the amplitude of the suspended mass.

Referring to Fig. 5.3(a), the ratio of the amplitudes for this case of resonance $\omega = \omega_n = 10$ Hz $= 20\pi$ rad/s is $\frac{Y}{X} = 2\zeta$. Also, logarithmic decrement $\delta = \frac{1}{N}\log_\varepsilon\left(\frac{x_1}{x_{N+1}}\right) = \frac{2\pi\zeta}{\sqrt{(1-\zeta^2)}} \doteqdot 2\pi\zeta$ for very small values of ζ. The amplitude ratio $\frac{x_1}{x_{N+1}} = 2$ in 2s, i.e. in 20 cycles or periods, i.e. $N = 5$.

Hence, $\frac{1}{20}\log_\varepsilon 2 \doteqdot 2\pi\zeta$ and $\zeta = \frac{0.6932}{40\pi}$ by means of which we estimate the amplitude of the suspended mass to be $X = \frac{Y}{2\zeta} = \frac{20\pi}{0.693}Y = 91\,Y$.

EXAMPLE 5.3(ii)—illustrations of resonance of a trailer pulled on an undulating road
A spring-mounted two-wheel trailer of mass M with negligible damping is towed at a speed V over an undulating road surface whose contour may be assumed to follow the law $y = Y\cos(\omega t + \beta)$ having a wavelength of undulations L. Estimate the forward speed of the trailer which will cause resonance, assuming harmonic response $x = X\cos\omega t$. If $M = 400$ kg and $k = 40$ N/mm and $L = 10$ m, calculate the forward

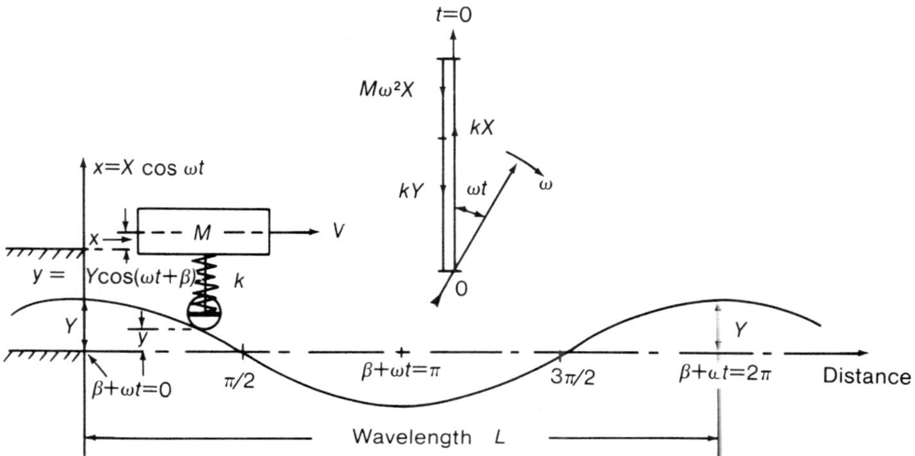

Fig. 5.3(c) Trailer on an undulating road

speed for resonance of the trailer, and the ratio of the amplitudes at speeds of 24 and 96 km/h.

In the position shown in Fig. 5.3(c) the springs, of total stiffness k, have each been stretched a distance $(x-y)$ and, therefore, exert a downward force on the mass of $k(x-y)$. Hence, $M\ddot{x} = -k(x-y)$ or $\ddot{x} + \omega_n^2 x = \omega_n^2 y$, as in Section 5.3. A but with $c = 0$.

The phase diagram of Fig. 5.3(c) shows that, without damping, $\beta = 0$ and $Y/X = 1 - M/k\omega^2 = 1 - \left(\dfrac{\omega}{\omega_n}\right)^2$. Resonance occurs when $\omega = \omega_n$, i.e. when $2\pi V/L = \sqrt{(k/M)}$ or when the forward velocity of the trailer is $V = \dfrac{L}{2\pi}\sqrt{(k/M)}$. Hence, in the particular case,

$$V = \frac{10\,\text{m}}{2\pi}\sqrt{\left(\frac{4\times 10^4\,\text{N}}{400\,\text{kg m}}\left[\frac{\text{kg m}}{\text{N s}^2}\right]\right)} = \frac{100\,\text{m}}{2\pi\ \text{s}} = 57.3\,\text{km/h},$$

and

$$\omega_n = 2\pi\frac{V}{L} = \sqrt{(k/M)} = \sqrt{\left(\frac{4\times 10^4\,\text{N}}{400\,\text{kg m}}\right)} = 10\frac{\text{rad}}{\text{s}}\left[\frac{\text{cycle}}{2\pi\,\text{rad}}\right] = 1.59\,\text{cycle/s}.$$

At $V = 24\,\text{km/h}$, $\omega = \dfrac{2\pi}{10}\times\dfrac{24}{3.6\text{s}} = 4.18/\text{s}$ and $\dfrac{\omega}{\omega_n} = 0.418$ and $\dfrac{X}{Y} = \dfrac{1}{1-(0.418)^2} = 1.21$.

At $V = 96\,\text{km/k}$, $\omega = 16.72/\text{s}$ and $\dfrac{\omega}{\omega_n} = 1.672$ and $\dfrac{X}{Y} = \dfrac{1}{1-(1.672)^2} = -1.82$ since $M\omega^2$ is greater than k.

(B) Consider the case when the base—say, the deck of a ship, moves (relative to the earth) according to $y = Y\cos\omega t$, and a spring-supported mass with

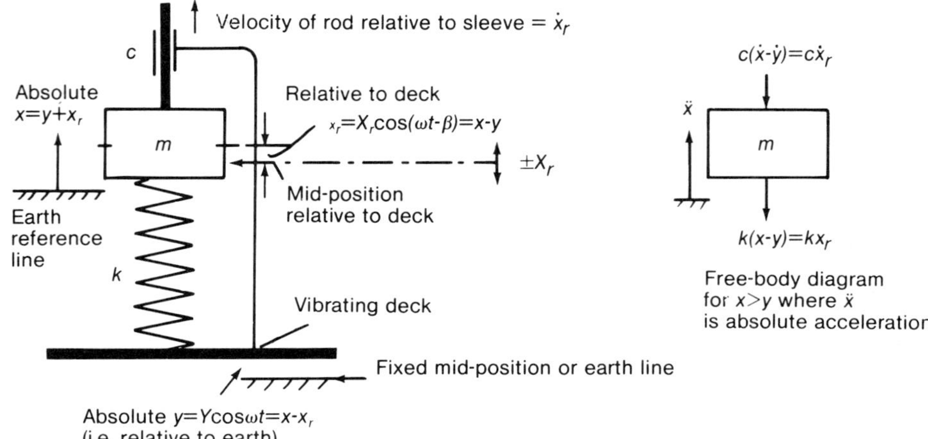

Fig. 5.3(d) Damped forced oscillations caused by moving base

viscous damping responds at the forcing frequency ω with harmonic motion $x_r = X_r \cos(\omega t - \beta)$ *relative to the moving base* as represented in Fig. 5.3(d). Since the mass m moves, from a fixed earth-datum line, an absolute distance x equal to the sum of an absolute distance y of a moving platform and the distance $(x - y)$ or x_r relative to that platform, we see that the mass has moved *upwards* from the earth-datum or fixed reference line a distance $x = y + x_r$ which is the absolute displacement of m; and the resisting force is $kx_r + c\dot{x}_r$ as indicated on the free-body diagram of Fig. 5.3(d).

Hence, the continuing or steady-state equation of motion is $m\ddot{x} = m(\ddot{y} + \ddot{x}_r) = -(kx_r + c\dot{x}_r)$ or $m\ddot{x}_r + c\dot{x}_r + kx_r = -m\ddot{y}$ which, on substituting, becomes $mY\omega^2 \cos\omega t = mX_r\omega^2 \cos(\omega t - \beta + \pi) + cX_r\omega \cos(\omega t - \beta + \pi/2) + kX_r \cos(\omega t - \beta)$ and is represented pictorially by the vector diagram on Fig. 5.3(e).

From the non-dimensional diagram of force ratios we deduce that the maximum displacement, X_r, of the mass relative to the deck divided by the maximum displacement, Y, of the deck relative to the earth, is given by

$$\frac{X_r}{Y} = \frac{(\omega/\omega_n)^2}{\sqrt{(\{1-(\omega/\omega_n)^2\}^2 + \{2\zeta\omega/\omega_n\}^2)}},$$

and the phase angle (or angle of lag), β, is given by

$$\tan\beta = \frac{2\zeta\omega/\omega_n}{1-(\omega/\omega_n)^2}.$$

The non-dimensional graphs of these are similar to those shown on Figs. 4.4

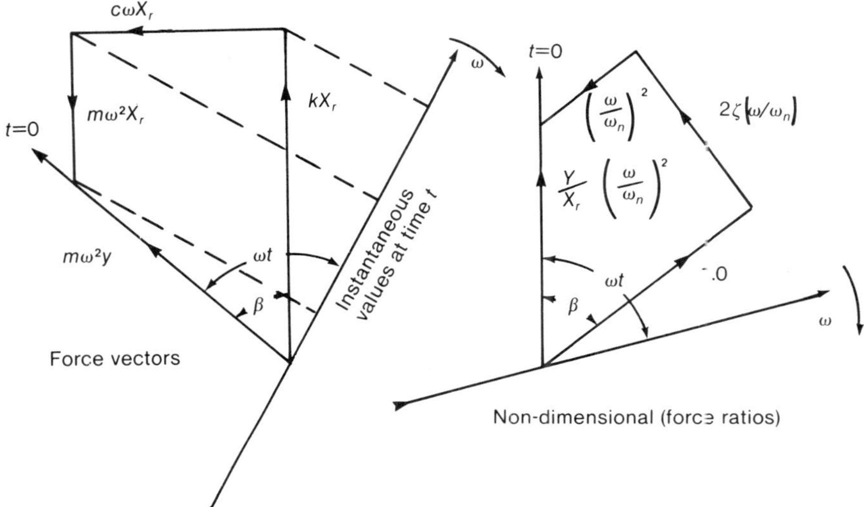

Fig. 5.3(e) Force and phase diagrams

(f and g), except that when ω/ω_n is high X_r/Y approaches 1.0. In practice it is often necessary to record the vibrations of, say, floors and decks, frameworks and structures, earthquakes and tremors, etc., in cases where there is no rigid base on which to mount a recording instrument. The nearest approach to a fixed base is that of supporting a large mass on springs of low-stiffness in order that the mass will have a natural frequency ($\sqrt{(k/m)}$) much lower than that of the vibration to be recorded. Thus, from a physical point of view, the vibrations or oscillations are relatively so rapid that the mass virtually stays fixed in space while the floor or deck vibrates. The basis of this type of measuring instrument is represented diagrammatically in Fig. 5.3(d) which when placed on a vibrating body determines the quantity to be measured via the *relative* motion between the seismic mass and the vibrating body. Figs. 5.3 (d and e) and resulting equations apply both to vibrometers (i.e. amplitude recorders) and accelerometers—the springs stiffness in the former being low compared with that of the latter.

In the vibrometer $\omega_n = \sqrt{(k/m)}$ is low (e.g. for engine work 5 Hz to 20 Hz) and, operating in a range where ω/ω_n is large, the non-dimensional graph shows that $\frac{X_r}{Y}$ approaches 1.0, i.e. the relative amplitude, X_r, is equal to the amplitude, Y, of the vibration being measured—the seismic mass remaining almost stationary at high values of ω.

In the accelerometer $\omega_n = \sqrt{(k/m)}$ is high relative to ω and, operating in a range where ω/ω_n is small with ζ approaching zero, $\frac{X_r}{Y}$ approaches $\left(\frac{\omega}{\omega_n}\right)^2$, i.e.

X_r is proportional to $\omega^2 Y$ which is the maximum acceleration to be measured in $\ddot{y} = -\omega^2 Y \cos \omega t$. If there is damping, the particular value of ζ which makes the denominator of the X_r/Y equation unity is given by

$$\left\{1 - \left(\frac{\omega}{\omega_n}\right)^2\right\}^2 + \left\{2\zeta\frac{\omega}{\omega_n}\right\}^2 = 1.0, \text{ i.e. } \zeta = \frac{1}{2}\sqrt{\left\{2 - \left(\frac{\omega}{\omega_n}\right)^2\right\}}$$

by means of which an approximate value of ζ may be found to suit a given range of ω/ω_n; e.g. if the operating range of the accelerometer is to be $\omega/\omega_n = 0$ to 0.5, the denominator may be made 1.0 for, say, $\omega/\omega_n = 0.3$ with $\zeta = 0.5\sqrt{(2 - 0.09)} = 0.69$. Constant damping can be obtained electrically, and the use of strain gauges for both static and dynamic tests of structures is often more appropriate and accurate.

EXAMPLE 5.3(iii)—illustrative of the limitations of a vibration pick-up
A vibration pick-up has a damping factor $\zeta = 0.6$ and a natural frequency of 5 cycles/s. Estimate the lowest frequency that can be measured with one per cent accuracy.

Diagrammatically, the pick-up can be represented by Fig. 5.3(f) which is a particular case of analysis (B). Referring to the phase diagram of Fig. 5.3(e), we have

$$\frac{X_r}{Y} = \frac{(\omega/\omega_n)^2}{\sqrt{\left(\left\{1 - \left(\frac{\omega}{\omega_n}\right)^2\right\}^2 + \left\{2\zeta\frac{\omega}{\omega_n}\right\}^2\right)}}$$

and, if $\frac{X_r}{Y} = 1.01$, then $1 - 2\left(\frac{\omega}{\omega_n}\right)^2 + \left(\frac{\omega}{\omega_n}\right)^4 + 1.44\left(\frac{\omega}{\omega_n}\right)^2 = 0.981\left(\frac{\omega}{\omega_n}\right)^4$

i.e. $\left(\frac{\omega}{\omega_n}\right)^4 - 29.5\left(\frac{\omega}{\omega_n}\right)^2 + 52.6 = 0$ and $\left(\frac{\omega}{\omega_n}\right) = 5.25$ or 1.38. Hence, the lowest frequency that can be measured with 1 per cent accuracy is 1.38×5 cycles/s = 6.9 Hz.

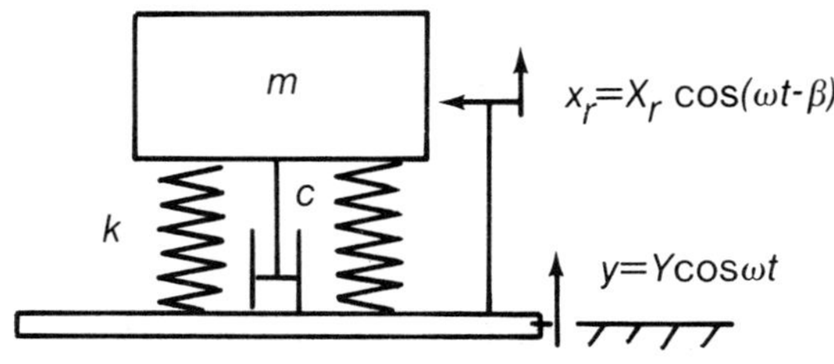

Fig. 5.3(f) Vibration pick-up

(C) Consider a mass suspended on a spring with viscous damping provided by lubrication between the mass and its oscillating guide-sleeve, as in Fig. 5.3(g).

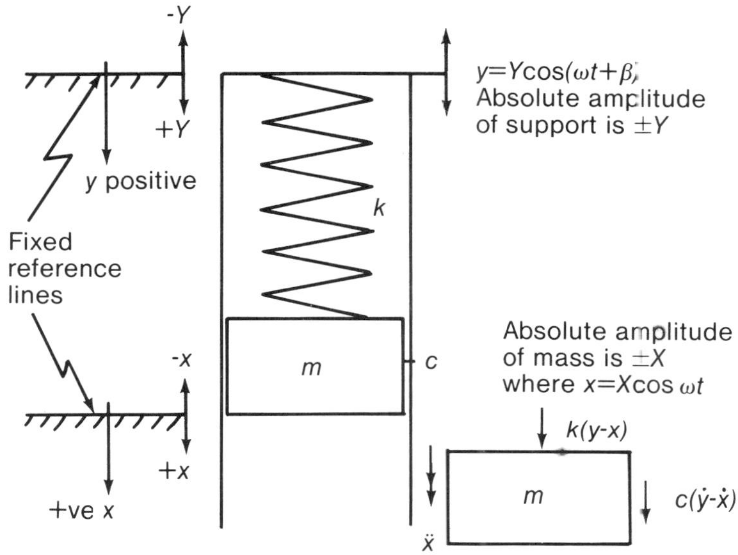

Fig. 5.3(g) Damped forced oscillations caused by a moving support-sleeve

Assume the support moves in vertical linear harmonic motion and that the mass responds within the oscillating sleeve with simple harmonic motion of the same frequency ω. The problem is to find the relationship (say, amplitude ratio $\dfrac{X}{Y}$ and phase) between the *absolute movements* of the support-sleeve and the mass—i.e. relative to the earth as distinct from relative to the moving base, which was the case in (B) and Fig. 5.3(d).

Referring to Fig. 5.3(g), let the instantaneous values of the displacements of the support and mass be y and x, respectively—both being measured from the mid-positions of their oscillations relative to the earth. The effective forces acting on the mass are $k(y - x)$ due to compression of the spring, and $c(\dot{y} - \dot{x})$ due to viscous drag—both of which act in the same direction because the velocity of m relative to the sleeve is $(\dot{y} - \dot{x})$.

Hence, $m\ddot{x} = c(\dot{y} - \dot{x}) + k(y - x)$ or $m\ddot{x} + c\dot{x} + kx = c\dot{y} + ky$, i.e. $m\omega^2 X \cos(\omega t + \pi) + c\omega X \cos\left(\omega t + \dfrac{\pi}{2}\right) + kX \cos \omega t = c\omega Y \cos\left(\omega t + \dfrac{\pi}{2} + \beta\right) + kY \cos(\omega t + \beta)$. This latter equation can be represented by the vector diagram shown in Fig. 5.3(h)—the maximum values of the three terms on the left of the equation being represented by vectors OA, AB and BD (as in Section 4.4), and the maximum values of the two terms on the right side of the equation by vectors OE and ED at right angles to each other. The amplitude and phase relationship can be found from the geometry of this vector diagram. For

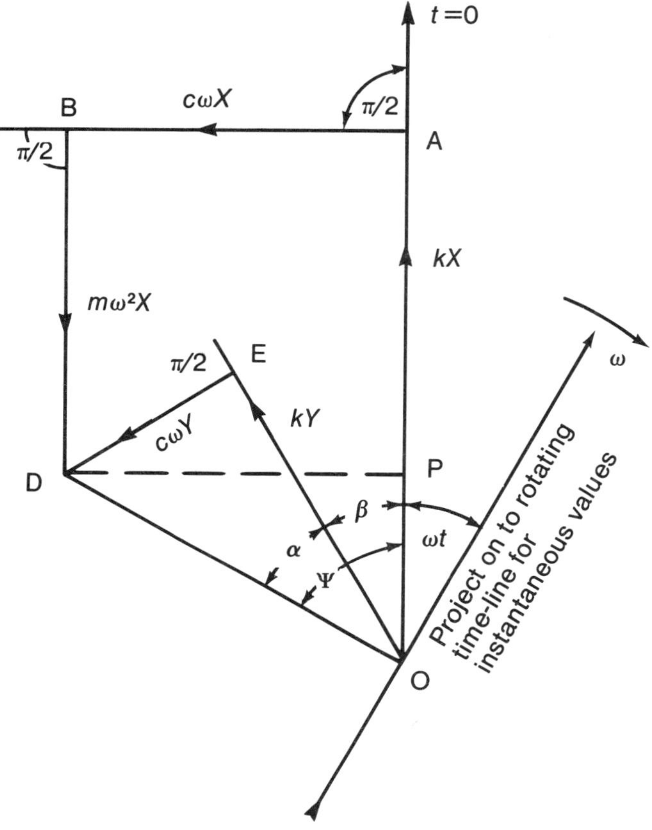

Fig. 5.3(h) Vector diagram of forces

example, since lengths are related by $OD^2 = OP^2 + DP^2 = OE^2 + ED^2$, we may write

$$\{(k - m\omega^2)^2 + (c\omega)^2\}X^2 = \{k^2 + (c\omega)^2\}Y^2$$

which, on dividing through by k^2 and using the previous relationships $\omega_n = \sqrt{\left(\dfrac{k}{m}\right)}$, $\zeta = \dfrac{c}{c_c} = \dfrac{c}{2\sqrt{(mk)}}$ —i.e. $\dfrac{2\zeta}{\omega_n} = \dfrac{c}{k}$ and $\left(\dfrac{\omega}{\omega_n}\right)^2 = \dfrac{m\omega^2}{k}$, results in the non-dimensional equation for the ratio of the maximum (absolute) amplitudes

$$\frac{X}{Y} = \sqrt{\left(\frac{1 + \left(2\zeta\dfrac{\omega}{\omega_n}\right)^2}{\left\{1 - \left(\dfrac{\omega}{\omega_n}\right)^2\right\}^2 + \left\{2\zeta\dfrac{\omega}{\omega_n}\right\}^2}\right)}$$

which is the same as the transmissibility equation of Section 4.6. Hence, a plot of maximum amplitude ratio, $\frac{X}{Y}$, against frequency ratio, $\frac{\omega}{\omega_n}$, gives the same family of curves as is shown on Fig. 4.6(b) all of which pass through the same point $\frac{\omega}{\omega_n} = \sqrt{2} = 1.414$, where $\frac{X}{Y} = 1.0$ and, after which, higher values of ζ give larger amplitudes than those for lower values of ζ.

The phase lag, β, between the maximum displacement, Y, of the disturbance and the maximum displacement, X, of the responding mass can be found from the fact that $\beta = \psi - \alpha$ where

$$\tan\psi = \frac{c\omega}{k - m\omega^2} = \frac{2\zeta\dfrac{\omega}{\omega_n}}{1 - \left(\dfrac{\omega}{\omega_n}\right)^2} \quad \text{and} \quad \tan\alpha = \frac{c\omega}{k} = 2\zeta\frac{\omega}{\omega_n}.$$

A practical case of this analysis is that of a vehicle travelling over a rough road. In the ideal design the mountings should be such that the body is not displaced owing to the roughness of the road surface. Although a vehicle has six degrees of freedom, the most important are vertical motion, pitching and rolling motion, as for a ship. Referring to Fig. 5.3(i) the spring-dashpot

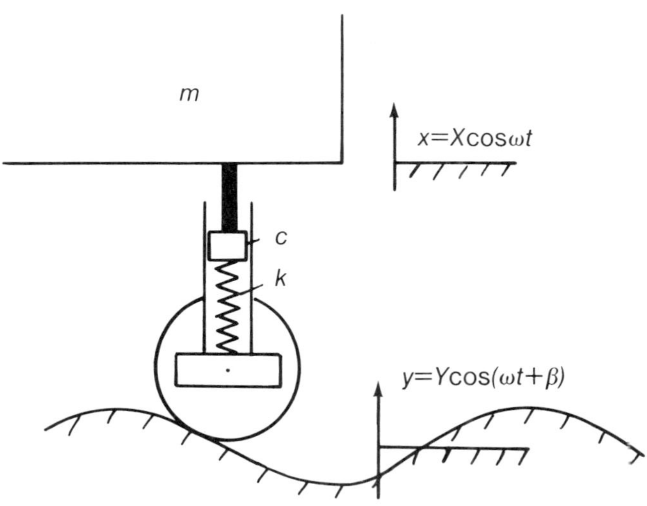

Fig. 5.3(i) Shock-absorbers for vehicles

shock-absorber is in effect the same as Fig. 5.3(g) upside down. Hence,

$$\frac{X}{Y} = \sqrt{\left(\frac{1 + \left(2\zeta \dfrac{\omega}{\omega_n}\right)^2}{\left\{1 - \left(\dfrac{\omega}{\omega_n}\right)^2\right\}^2 + \left\{2\zeta \dfrac{\omega}{\omega_n}\right\}^2} \right)}$$

where $\omega_n = \sqrt{\left(\frac{k}{m}\right)}$ and ω is the angular frequency of the assumed harmonic motion of the vertical displacement of the wheel caused by undulations or roughness of the road surface. If the springs are farily 'soft', ω_n is relatively low and if $\frac{\omega}{\omega_n}$ is high we see from Fig. 4.6(b) that $\frac{X}{Y}$ will be less than 1.0 and lower the lower the values of ζ. Thus, a smooth ride will result if the springs are soft and the damping coefficient less than about half the critical, i.e. $\zeta = 0.5$.

EXAMPLE 5.3(iv)—illustrative of the value of clear diagrams incorporating the essentials of a vibrating system

Draw a diagram of a damped system mounted on a vibrating base and from which the fundamental equations of motion of the two cases analysed in paragraphs B & C can be clearly deduced.

Taking the upwards vertical to be the positive direction of linear displacement, the diagrammatic sketch of Fig. 5.3(j) is best drawn so as to show positive displacements of y, x_e and x_b, otherwise there is likely to be difficulty with signs. Distances y and x_e are

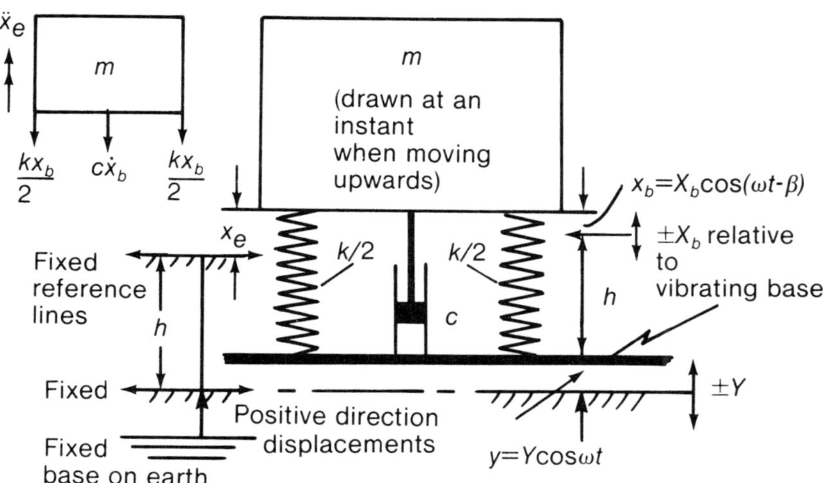

Fig. 5.3(j) A damped system on a vibrating base

measured from reference lines fixed on the earth, and x_b from a reference line vibrating with the base.

Since $h + x_e = y + h + x_b$ or $x_e = y + x_b$ and the springs are extended by x_b, the force exerted on the mass by the springs is kx_b downwards, where k is the total stiffness; and the viscous force exerted on the mass is $c\dot{x}_b$ downwards. The acceleration of the mass is $\ddot{x}_e$ upwards. Hence, applying the second law of motion, we have

$$m\ddot{x}_e = -(kx_b + c\dot{x}_b) = m(\ddot{y} + \ddot{x}_b)$$

i.e. $m\ddot{x}_b + c\dot{x}_b + kx_b = -m\ddot{y}$, which is the equation of motion in which x_b is measured from a reference fine (or pen) vibrating with the base, as in sub-section B.

Alternatively, since $x_b = x_e - y$, we may write

$$m\ddot{x}_e = -\{k(x_e - y) + c(\dot{x}_e - \dot{y})\}$$

i.e. $m\ddot{x}_e + c\dot{x}_e + kx_e = c\dot{y} + ky$, which is the equation of motion in which x_e is measured from a reference line fixed to the earth, as in sub-section C.

EXAMPLE 5.3(v)—illustrative of models of systems with several degrees-of-freedom e.g. an automobile

Set up mathematical models of an automobile for (a) analysis of vertical motion and (b) for vertical and pitching motion.

(a) Representing the multiple-degree-of-freedom vibration system of an automobile as a 'lumped parameter model', there are three elements of mass, namely (i) m_u,

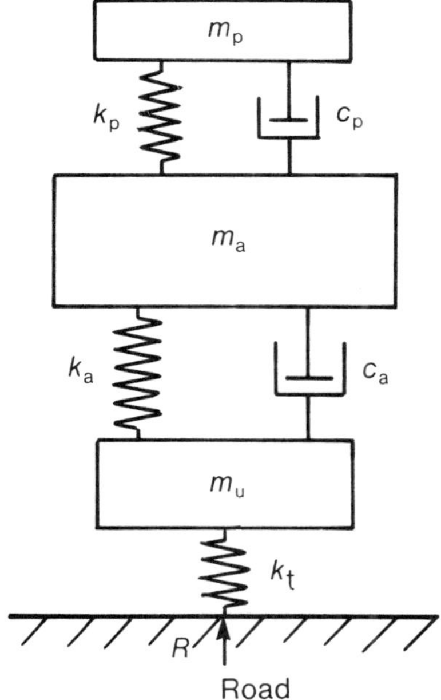

Fig. 5.3(k) Vertical motion of automobile model

unspring except for types of elastic stiffness k_t (ii) m_a, of the chassis supported by the main springs of stiffness k_a and shock-absorbers of damping coefficient c_a and (iii) m_p, of the seats and passengers on springs of stiffness k_p and damping coefficient c_p.

Fig. 5.3(k) represents the mathematical model of minimum complexity for evaluating the quantities concerning the riding comfort of passengers. A variety of motions representing road irregularities can be applied at R and the effect on m_p of the stiffness (k_t) of tyres, stiffness (k_a) and damping (c_a) of main suspension and (k_p, c_p) of the seat assembly may be estimated.

(b) Taking pitching—i.e. front moving upwards while the rear moves downwards or vice versa, into account by devising a model with two complete wheel and tyre assemblies suffering road irregularities R_F and R_R, Fig. 5.3(1) shows the model which represents the riding characteristics of an automobile more realistically. However, the complexity of the analysis increases in consequence. Thus, the chassis of mass m_a is considered to be a rigid body free to move vertically and in angular motion. Hence, one differential equation is needed for the vertical linear motion of each lumped mass, m_u, of each axle and wheel-pair, one for vertical motion, and one for angular motion of chassis m_a—i.e. a five degrees-of-freedom system before arriving at the passenger-seating system within the vehicle. Thus, in addition to analytical results for vertical motion would be an expression giving the angular position of the chassis as a function of time. Further analysis of this necessitates the writing of a program for use in a computer. See Appendix D, (10) and (11).

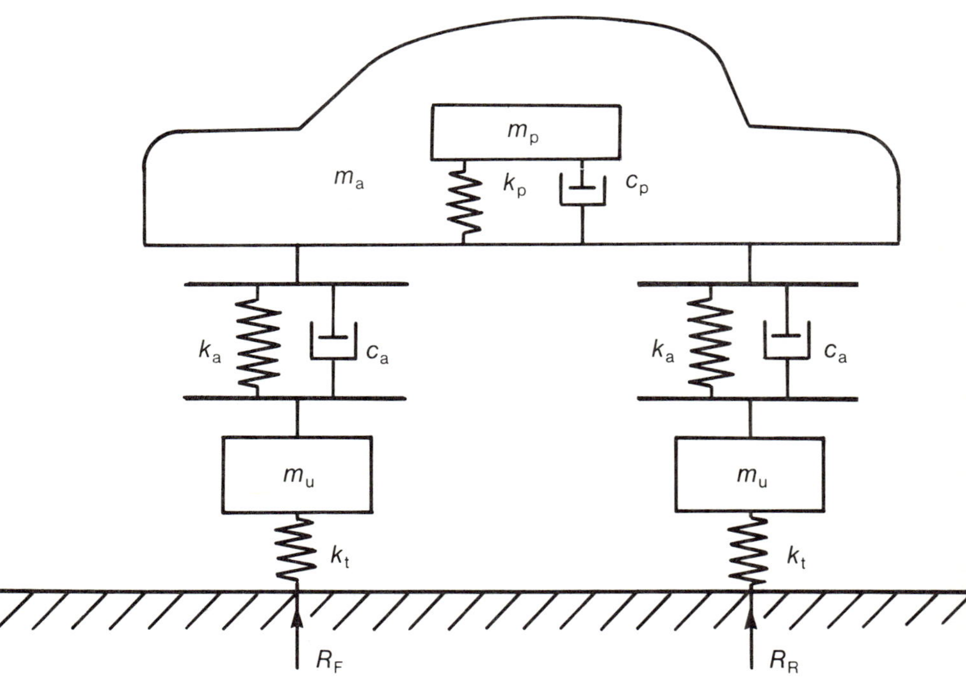

Fig. 5.3(l) Model for analysis of vertical and pitching motions

EXERCISES ON CHAPTER 5

1. The springs of an automobile trailer are compressed 100 mm under its weight. Neglecting damping, calculate the natural frequency and forward speed which will cause resonance when the trailer is travelling over an undulating road assumed to have the profile of a harmonic of wavelength 15 m and amplitude 75 mm. Also, estimate the amplitudes of the vibration of the trailer when travelling at 60 km/h and 100 km/h.
2. For the simple vibration neutralizer with $\omega_{n1} = \omega_{n2}$ and a mass ratio $\mu = m_2/m_1$ with upper and lower natural frequency ratios $r_1 = \omega_1/\omega_{n2}$ and $r_2 = \omega_2/\omega_{n2}$, show that $r_{1,2} = (1 + \mu/2) \pm \sqrt{\{(1 + \mu/2)^2 - 1\}}$ or $r_1^2 r_2^2 = 1$ and $r_1^2 + r_2^2 = 2 + u$.
3. A recording instrument is found to have a natural damped frequency of 99.4 cycles/min and, when left to itself its amplitude is halved in one complete swing or cycle. Calculate the damping factor ζ and the undamped natural frequency f_n. (b) If this instrument is used to record a harmonic force of frequency 80 cycles/min, estimate the relative magnitudes and phase relations of the force and the record made by the instrument, i.e. $X/F/k$ and β.
4. (a) In a reed-type frequency meter connected to a 50 Hz supply, the amplitude of the reed whose natural frequency is 50 Hz is 6.3 mm and that of a 60 Hz reed is 0.25 mm. Assuming viscous damping and that k and ζ are the same for both reeds, also that the applied force is the same at all frequencies, estimate the value of ζ. (b) If the frequency is reduced to 49.5 Hz, calculate the amplitude of the 50 Hz reed and its semi-life when the supply is switched off.
5. A mass m is suspended by a helical spring which hangs froma support vibrating in s.h.m. with amplitude Y at the natural frequency of the spring-mass system. To prevent excessive oscillations due to resonance, a dashpot is introduced (see Fig. 5.3(a)) to provide viscous damping. Prove that in order to limit the amplitude of the oscillations of m to X, the value of the damping coefficient, c, must not be less than $\omega m Y/X$.

6

Continuous-mass systems

6.1 TRANSVERSE AND LONGITUDINAL VIBRATIONS

In free undamped or autonomous oscillations the only external forces applied are those necessary to keep the ends or pivots fixed. If, however, a periodic force were applied, e.g. if the end of a rod or thin beam is made to oscillate in a particular way, this would induce a motion of the same period in the rest of the material and (as with induced linear harmonic oscillations of Section 5.3) such induced oscillations are in the category of forced oscillations.

One important difference between free and forced vibrations is that, although there are an infinite number of possible free or undamped natural oscillations, the frequency of a forced oscillation is always the same as that of the disturbing force and can, therefore, have any magnitude. Hence, it is important to know the lower natural frequencies of beams, e.g. in bridges, aircraft wings and beams supporting vibrating machines, so as to know when resonance and consequent large bending stresses are likely to occur.

Beams and plates are continuous systems with distributed mass and stiffness properties. The critical or whirling speed of a rotating shaft is a special case of the natural frequency of such systems. Thus, considering the flexural vibration of beams in reference to Fig. 6.1(a), we deduce that if x is constant, the position of any point P varies with time, t, and that at any particular instant of time the position, y, of P varies with x. Hence, $y = f(x, t)$ (see Appendix B10 (iv)) and, for any particular x, the acceleration of the element δx is $\dfrac{\partial^2 y}{\partial t^2}$. Therefore, for the element δx of a beam of cross-sectional area a and second moment of area I, the force in the Y direction on the element of mass $\rho a \delta x$ is $\delta F = (\rho a \delta x)\dfrac{\partial^2 y}{\partial t^2}$. Hence, for linear transverse motion

$$\rho a \frac{\partial^2 y}{\partial t^2} = \frac{\partial F}{\partial x} = -\frac{\partial^2 M}{\partial x^2} = -EI\frac{\partial^4 y}{\partial x^4},$$

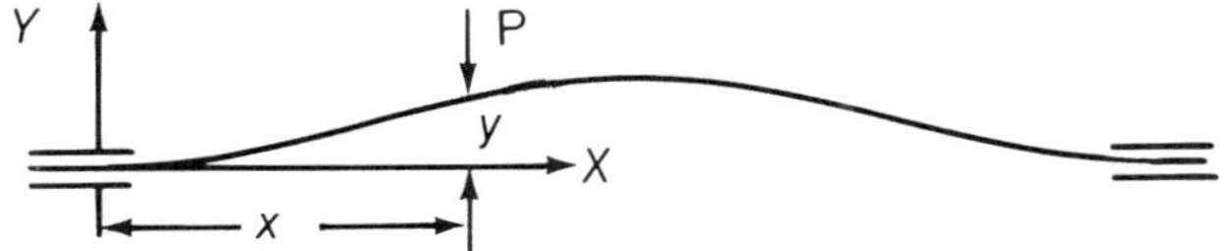

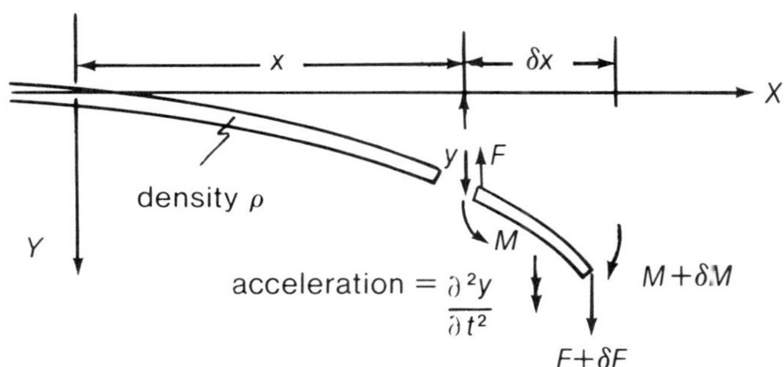

Fig. 6.1(a) Transverse vibrations

where E is the modulus of elasticity of the material (see Appendix A4), and the slope $\frac{dy}{dx}$ is small. This fourth-order partial differential equation

$$EI\frac{\partial^4 y}{\partial x^4} = -\rho a\frac{\partial^2 y}{\partial t^2} = -\frac{w}{g}\frac{\partial^2 y}{\partial t^2},$$

where w is the weight intensity per unit length of beam, applies to all cases of free transverse vibration of beams whether loaded or vibrating because of their own masses only. The formula implies that the lateral oscillations are small so that rotational motion is negligible, and the simple bending theory yielding

$$\frac{1}{R} = \frac{\pm\dfrac{d^2y}{dx^2}}{\left\{1+\left(\dfrac{dy}{dx}\right)^2\right\}^{3/2}} = \frac{M}{EI} \doteq \frac{d^2y}{dx^2}$$

(see Appendix A4) applies. This is based on the assumptions that plane cross-sections of a beam remain plane during flexure, and that the radius of curvature of a bent beam is large compared with the depth of the beam.

In vibration analysis of deep beams, both rotary inertia and shear deformation must be taken into account, and the kinetic energy of rotation of cross-sectional planes about the neutral axis when passing through the static-equilibrium position will be a significant proportion.

In the case of forced lateral vibration of thin beams, the equation of motion of an element of the beam for non-autonomous vibration is

$$EI\frac{\partial^4 y}{\partial x^4} + \rho a\frac{\partial^2 y}{\partial t^2} = f(x, t),$$

where $f(x, t)$ represents the time-varying force distribution (i.e. an applied dynamic load/unit length) acting on the beam in the y direction.

A continuous system is the limit of a 'lumped-mass' system in which the distance between any two adjacent mass points approaches zero and in which, in a wire, rod or thin beam, there are an infinite number of mass points. Consequently, a continuous system has an infinite number of degrees-of-freedom because an infinite number of displacements, $f(x, t)$, must be ascertained in order to describe the configuration of the system. Thus, whereas a lumped-mass system with n degrees-of-freedom requires only n ordinary differential equations, a continuous system, with an infinite number of degrees-of-freedom, requires one partial differential equation, i.e. the wave equation which governs the dynamics. Flexural autonomous vibration (earlier in this Section and Section 6.2) is governed by a fourth order wave equation, whereas longitudinal and torsional autonomous vibration (see Section 6.3) in an elastic shaft have a second order wave equation.

EXAMPLE 6.1(i)—illustrative of longitudinal vibration

Deduce the wave equation governing longitudinal autonomous vibration of an elastic bar or shaft, and state why this continuous elastic member has an infinite number of degrees of freedom.

Referring to Fig. 6.1(b) and assuming constant cross-sectional area A, homogeneous (constant ρ) isotropic (constant E) material, and that each cross-sectional plane remains a plane as it moves longitudinally by a distance $y = f(x, t)$ (also, see Appendix B10(iv)), let the instantaneous displacement of the plane at x be y, then at $x + \delta x$ it is $y + \delta y$.

The normal strain in direction x caused by displacement y is $\varepsilon = \dfrac{\partial y}{\partial x}$ caused by normal stress $\sigma = E\varepsilon$ which, in turn, is caused by a normal force $F = A\sigma = AE\dfrac{\partial y}{\partial x}$. Similarly, at section $x + \delta x$, the force is

$$F + \delta F = AE\varepsilon_{x+\delta x} = AE\left(\frac{\partial y}{\partial x} + \frac{\partial}{\partial x}\left(\frac{\partial y}{\partial x}\right)\delta x\right) = AE\left(\frac{\partial y}{\partial x} + \frac{\partial^2 y}{\partial x^2}\delta x\right).$$

Hence, the resultant force δF causing acceleration of δm in direction x is $\delta F = (A\rho\delta x)\dfrac{\partial^2 y}{\partial t^2} = AE\dfrac{\partial^2 y}{\partial x^2}\delta x$, i.e. $\dfrac{\partial^2 y}{\partial x^2} = \dfrac{\rho}{E}\dfrac{\partial^2 y}{\partial t^2}$ which is a linear partial differential equation

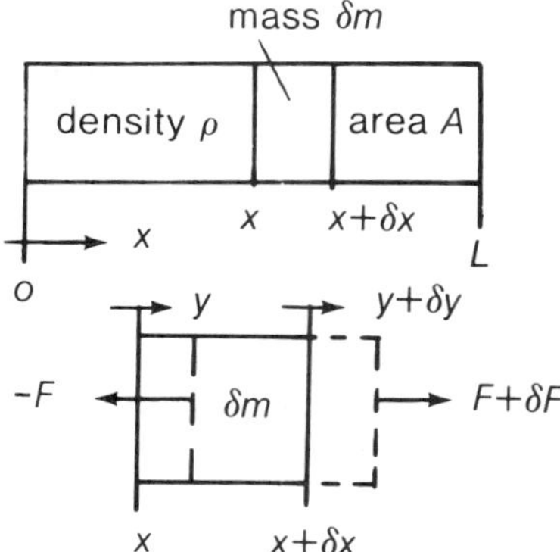

Fig. 6.1(b) Uniform elastic bar

with constant coefficients, and is the wave equation governing the autonomous longitudinal vibrations.

Since we must specify, at any instant of time, the displacement y of every plane in order to describe the instantaneous configuration of the system, and there are an infinite number of planes in a finite length of bar, the number of degrees of freedom of this continuous elastic member is infinite.

6.2 NORMAL MODES OF TRANSVERSE OSCILLATIONS

A rod or thin beam can oscillate transversely in an infinite number of ways, in every one of which the displacement satisfies the fundamental equation $EI\dfrac{\partial^4 y}{\partial x^4} = -\rho a\dfrac{\partial^2 y}{\partial t^2}$. There are, however, certain types or modes of oscillation in each of which every particle of the rod executes simple harmonic motion in the same period and same phase, but of different amplitudes—i.e. the periodic time is independent of x but the amplitude is a function of x. These modes are called *normal modes* and are expressed by $y = u\sin(\omega t + \alpha)$ where u, the amplitude, is a function of x and not of t, as is shown by inserting y in the fundamental equation to give

$$EI\sin(\omega t + \alpha)\frac{\partial^4 u}{\partial x^4} + \rho a u\{-\omega^2\sin(\omega t + \alpha)\} = 0$$

or

$$\frac{\partial^4 u}{\partial x^2} - \frac{\rho a\omega^2 u}{EI} = [\mathrm{D}^4 - \lambda^4]u = 0, \quad \text{where } \lambda^4 = \frac{\rho a\omega^2}{EI}.$$

This latter differential equation does not involve t and, therefore, the assumption that $y = u \sin(\omega t + \alpha)$, where $u = \psi(x)$, is verified. Hence, we may deduce (see Appendix B12(v)) that since $[D^2 - \lambda^2][D^2 + \lambda^2]u = 0$ then $u = A \cos \lambda x + B \sin \lambda x + C \cosh \lambda x + D \sinh \lambda x$.

Fixings of the vibrating rod (or beam) and end conditions (e.g. (i) free or no restraint, (ii) pinned or lateral movement impossible, but no restraint on rotation, (iii) encastré or no lateral or rotation possible) enable the constants A, B, C and D to be determined (see Examples 6.2(i), (ii), (iii)).

EXAMPLE 6.2(i)—illustrative of transverse vibration

In the case of a rod of length l clamped at one end and free at the other $x = 0$, $y = 0$, i.e. $u = 0 = A + 0 + c + 0$ or $C = -A$. Also, $x = 0$, $\frac{dy}{dx} = 0$, i.e. $\frac{du}{dx} = 0 = 0 + \lambda B + 0 + \lambda D$ or $D = -B$. Hence, $u = A(\cos \lambda x - \cosh \lambda x) + B(\sin \lambda x - \sinh \lambda x)$. At the free end, $x = l$, $M = 0$, i.e. $\frac{\partial^2 y}{\partial x^2} = 0$ and $\frac{d^2 u}{dx^2} = 0 = A(\cos \lambda l + \cosh \lambda l) - B(\sinh \lambda l + \sinh \lambda l)$. Also, $x = l$, $F = 0$, i.e. $\frac{d^3 y}{dx^3} = 0 = A(\sin \lambda l - \sinh \lambda l) - B(\cos \lambda l + \cosh \lambda l)$. If the rod is vibrating, A and B are not zero. Hence,

$$(\cos \lambda l + \cosh \lambda l)^2 + (\sin^2 \lambda l - \sinh^2 \lambda l) = 0$$

$$= 2(1 + \cos \lambda l \cosh \lambda l), \text{ i.e. } \cos \lambda l \cosh \lambda l = -1.$$

This equation determines λl and, hence, $\omega^2 = \lambda^4 EI/\rho a$. As there are an infinite number of roots to $\cos \lambda l = -\operatorname{sech} \lambda l$ found from Fig. 6.2(a), namely, $\lambda l = 1.88, 3\pi/2, 5\pi/2 \ldots$, there are an infinite number of phase rates ω, each of period $2\pi/\omega$, where $\omega = \lambda^2 \sqrt{\left(\frac{EI}{\rho a}\right)} = \lambda^2 l^2 \sqrt{(EI/\rho a l^4)}$ which is $(1.88)^2 \sqrt{(EI/\rho a l^4)}$ for the fundamental mode.

EXAMPLE 6.2(ii)—illustrative of transverse vibration of cantilever beams

Assuming that the natural vibration of a cantilever of uniform section, encastré at one end and free at the other, is of a certain standard form in which the value of λl is given by

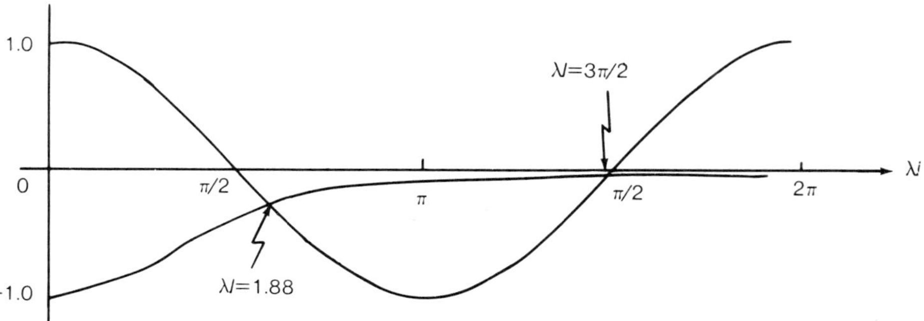

Fig. 6.2(a) Graphs of $\cos \lambda l$ and—$\operatorname{sech} \lambda l$

the equation $\cos \lambda l \cosh \lambda l + 1 = 0$, estimate the ratio between the frequency of natural vibration of a turbine blade of uniform section vibrating in its primary and secondary modes.

From the graph of $\cos \lambda l \cosh \lambda l = -1$ of Fig. 6.2(a) we find that λl is 0.6π and 1.49π for the primary and secondary modes of vibration, respectively. Hence, since $\omega = (\lambda l)^2 \sqrt{\left(\frac{EI}{\rho a l^4}\right)}$, the ratio of the secondary to primary frequencies is

$$\frac{\omega_2}{\omega_1} = \left(\frac{\lambda_2 l_2}{\lambda_1 l_1}\right)^2 = \left(\frac{1.49}{0.6}\right)^2 = 6.16.$$

This equation $\cos \lambda l \cosh \lambda l = -1$ could, of course, readily be solved using a root-finding algorithm on a micro-computer.

EXAMPLE 6.2(iii)—illustrative of transverse vibration of a loaded beam
If the vibrating rod of Example 6.2(i) has a weight W on the free end then, as before, $x = 0$, $y = 0$; and $x = 0$, $\frac{\partial y}{\partial x} = 0$. Hence, using $\lambda^4 = \frac{\rho a \omega^2}{EI}$, we have $u = A(\cos \lambda x - \cosh \lambda x) + B(\sin \lambda x - \sinh \lambda x)$. Also, at the free end, $x = l$, $M = 0 = \frac{\partial^2 y}{\partial x^2}$ and hence

$$\frac{d^2 u}{dx^2} = 0 = -\lambda^2 A(\cos \lambda l + \cosh \lambda l) - \lambda^2 B(\sin \lambda l + \sinh \lambda l) \tag{1}$$

but when $x = l$, the shearing force on the rod is $F_1 = -\frac{W}{g}\frac{\partial^2 y_1}{\partial t^2}$ in accordance with Fig. 6.2(b) and Appendix A4. Hence, since $\frac{\partial M}{\partial x} + F = 0$, it follows that at the free end

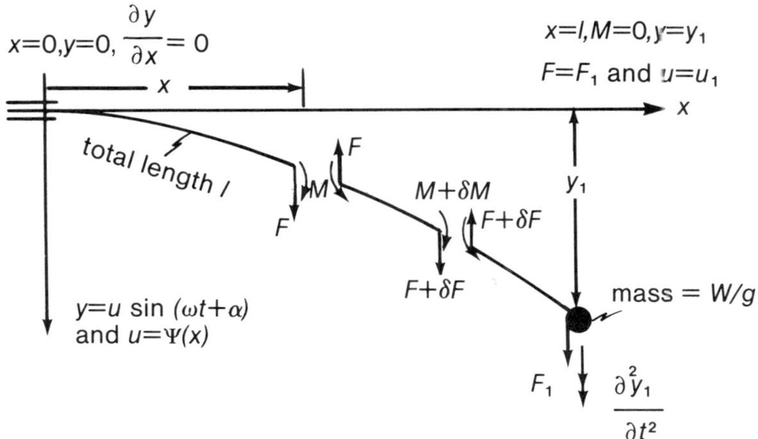

Fig. 6.2(b) Weighted vibrating cantilever

$$EI\frac{\partial^2 y_1}{\partial x^3} - \frac{W}{g}\frac{\partial^2 y_1}{\partial t^2} = 0 \text{ i.e. } EI\frac{\partial^3 u_1}{\partial x^3} + \frac{W}{g}\omega^2 u_1 = 0 \text{ or } \frac{\partial^3 u_1}{\partial x^3} = -\frac{W}{g}\frac{\lambda^4 u_1}{\rho a}$$

which expands into

$$\frac{\partial^3 u_1}{\partial x^3} = -\lambda^3 A(-\sin\lambda l + \sinh\lambda l) - \lambda^3 B(\cos\lambda l + \cosh\lambda l)$$

$$= \frac{-W\lambda^4}{g\rho a}\{A(\cos\lambda l - \cosh\lambda l) + B(\sin\lambda l - \sinh\lambda l)\} \qquad (2)$$

From equation (1) $\frac{A}{B} = -\left(\frac{\sin\lambda l + \sinh\lambda l}{\cos\lambda l + \cosh\lambda l}\right)$, and substituting this ratio in a rearranged equation (2), we get:

$$(1 + \cos\lambda l\cosh\lambda l) = \frac{W}{g\rho a}(\sin\lambda l\cosh\lambda l - \cos\lambda l\sinh\lambda l)$$

which can be graphed to find λl. Hence, $\omega^2 = \frac{\lambda^4 EI}{\rho a}$ and $\tau = \frac{2\pi}{\omega}$.

EXAMPLE 6.2(iv)—illustrative of transverse vibration

In the case of a rod or beam pinned at both ends as represented in Fig. 6.2(c), we have, as before, $y = u\sin(\omega t + \alpha)$ where u is a function of x, and $[D^4 - \lambda^4]u = 0$. Hence (see Appendix B12 (v)), $u = A\cos\lambda x + B\sin\lambda x + C\cosh\lambda x + D\sinh\lambda x$. Also, when $x = 0$, $y = 0$, $M = 0$, therefore $C = A = 0$. At the other end $x = l$, $y = 0$, $M = 0$, therefore $u = 0 = B\sin\lambda l + D\sinh\lambda l$ and $M = 0 = \frac{\partial^2 y}{\partial x^2}$, therefore $\frac{\partial^2 u}{\partial x^2} = 0 = -\lambda^2 B\sin\lambda l + \lambda^2 D\sinh\lambda l$. Hence, for a non-zero solution (i.e. B and $D \neq 0$), $\sin\lambda l\sinh\lambda l = 0$; but since $\lambda l \neq 0$, $\sinh\lambda l \neq 0$ and $\sin\lambda l = 0$, i.e. $\lambda l = 0, \pi, 2\pi, \ldots n\pi$. Also $(\lambda l)^2 = \omega l^2\sqrt{\left(\frac{\rho a}{EI}\right)} = \pi^2, 4\pi^2 \ldots n^2\pi^2$, and the periodic time of the nth mode is $\tau_n = \frac{2\pi}{n\omega} = \frac{2l^2}{\pi n^2}\sqrt{\left(\frac{\rho a}{EI}\right)}$.

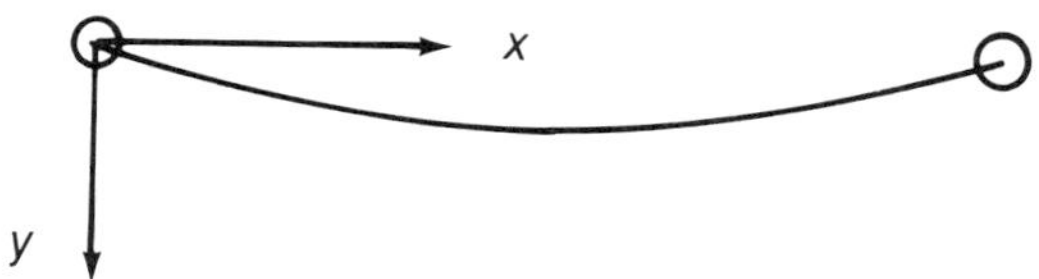

Fig. 6.2(c) Vibrating beam pinned at ends

Any free transverse vibration is usually a combination of normal modes. It is unlikely that a rod given a blow, or deflected and let go, will begin to vibrate in one of the normal modes unless the rod were flexed at the start in an extreme position of a normal mode. However, in the composition of a number of normal modes the first is the most important. For example, in the case of the bar pinned at both ends (Example 6.2(iii)) for which $B \sin \lambda l + D \sinh \lambda l = 0$ and $B \sin \lambda l - D \sinh \lambda l = 0$ i.e. $2D \sinh \lambda l = 0$ but, since $\lambda l \neq 0$, then $D = 0$. Also, $2B \sin \lambda l = 0$ but since A, C and $D = 0$ then $B \neq 0$ if there is to be a non-zero solution. Therefore, $u = B_n \sin \lambda x = B_n \sin \frac{n\pi}{l} x$ where n is any integer and, for the nth normal mode,

$$y_n = B_n \sin \frac{n\pi}{l} x \sin(\omega_n t + \alpha), \text{ where } \omega_n = \frac{n^2\pi^2}{l^2} \sqrt{\left(\frac{EI}{\rho a}\right)}.$$

Also,

$$\frac{\partial y_n}{\partial t} = B_n \omega_n \sin \frac{n\pi}{l} x \cos(\omega_n t + \alpha).$$

Thus, representing y as the sum of an infinity of normal modes we may write $y = y_1 + y_2 + y_3 + \cdots$ to ∞. At the beginning, $t = 0$ and $y = B_1 \sin \frac{\pi x}{l} \sin \alpha_1 + B_2 \sin 2\pi \frac{x}{l} \sin \alpha_2 + \cdots$ to ∞ and also when $t = 0$, $\frac{\partial y}{\partial t} = B_1\omega_1 \sin \pi \frac{x}{l} \cos \alpha_1 + B_2\omega_2 \sin 2\pi \frac{x}{l} \cos \alpha_2 + \cdots$ to ∞. The right hand sides of y and $\frac{\partial y}{\partial t}$ are Fourier series Y and it is (see Section 1.3) possible to determine the coefficients so as to make each of the series represent any given single-valued continuous function such as $F(x)$ and $f(x)$,—$F(x)$ being the shape of the curve when $t = 0$, and $f(x) = \frac{\partial y}{\partial t}$ being the velocity of a point. Thus, we may write $F(x) = B_1 \sin \alpha_1 \sin \pi \frac{x}{l} + B_2 \sin \alpha_2 \sin 2\pi \frac{x}{l} + \cdots$ and $f(x) = \omega_1 B_1 \cos \alpha_1 \sin \pi \frac{x}{l} + \omega_2 B_2 \cos \alpha_2 \sin 2\pi \frac{x}{l} + \cdots$. The nth coefficient of $\sin n\pi \frac{x}{l}$ is given by $B_n \sin \alpha_n = \frac{2}{l} \int_0^l F(x) \sin n\pi \frac{x}{l} \mathrm{d}x$ and, similarly, $\omega_n B_n \cos \alpha_n = \frac{2}{l} \int_0^l f(x) \sin n\pi \frac{x}{l} \mathrm{d}x$. Thus, the arbitrary constants B_n and α_n can be found and, by putting $n = 1, 2, 3 \ldots$ all the constants $B_1, B_2, B_3 \ldots$, then $\alpha_1, \alpha_2, \alpha_3 \ldots$ can be found. Hence, y is known completely at any time t and at any x. Complete vibratory motion is thus composed of an infinite number of oscillations of different periods, and requires computer-programming for its analysis.

6.3 NATURAL FREQUENCIES OF SHAFTS IN TORSIONAL OSCILLATION

Practical examples are long boring-bars and drill-tubes for mines and wells. A vibrating or oscillating shaft or tube is a continuous-mass or multi-mass system. It will be appreciated from Fig. 6.3(a) that the angle of twist along a torsionally-vibrating shaft of uniform cross-section is a function of length and time. Since $\frac{T}{J} = \frac{G\theta}{L}$, where J is the polar second moment of area of the cross-section, namely, $\frac{\pi}{2}r^4$, we may write for the element δx that

$$\delta\theta = \frac{T\delta x}{GJ} \quad \text{and} \quad \frac{\partial T}{\partial x} = GJ\frac{\partial^2\theta}{\partial x^2}.$$

Also, since $T = I\ddot{\theta}$ in general, where I is the moment of inertia, namely, $\rho\frac{\pi}{2}r^4\delta x$ for the element (i.e. $I = \rho J\delta x$), then the net torque on the element is $\delta T = (\rho J\delta x)\frac{\partial^2\theta}{\partial t^2}$ and, hence, $\frac{\partial^2\theta}{\partial t^2} = \left(\frac{G}{\rho}\right)\frac{\partial^2\theta}{\partial x^2}$—the general solution of which, for harmonic vibration ($\theta = f(x) \times \sin\omega t$) of every element, is obtained from

$$-\omega^2 f(x)\sin\omega t = \left(\frac{G}{\rho}\right)\sin\omega t\frac{\mathrm{d}^2f(x)}{\mathrm{d}x^2} \text{ or } \frac{\mathrm{d}^2f(x)}{\mathrm{d}x^2} + \frac{\rho}{G}\omega^2 f(x) = 0.$$

Hence, $f(x) = A\sin(\omega x\sqrt{\rho/G}) + B\cos(\omega x\sqrt{\rho/G})$ and the equation for angle of twist at any section x is

$$\theta = f(x)\sin\omega t = \sin\omega t\{A\sin(\omega x\sqrt{\rho/G}) + B\cos(\omega x\sqrt{\rho/G})\},$$

where A and B are constants of integration and depend on the boundary conditions of the particular problem, e.g. if one end of the shaft is fixed and the other free, the boundary conditions are $x = 0$, $\theta = 0$ and $x = L$, $T = 0 = \frac{\partial\theta}{\partial x}$, i.e. $B = 0$, and $\frac{\partial\theta}{\partial x} = 0 = \sin\omega t\{A\omega\sqrt{\rho/G}\cos(\omega L\sqrt{\rho/G})\}$ which is satisfied if $\omega L\sqrt{\rho/G} = \frac{\pi}{2}, \frac{3}{2}\pi, \ldots (n + 1/2)\pi$.

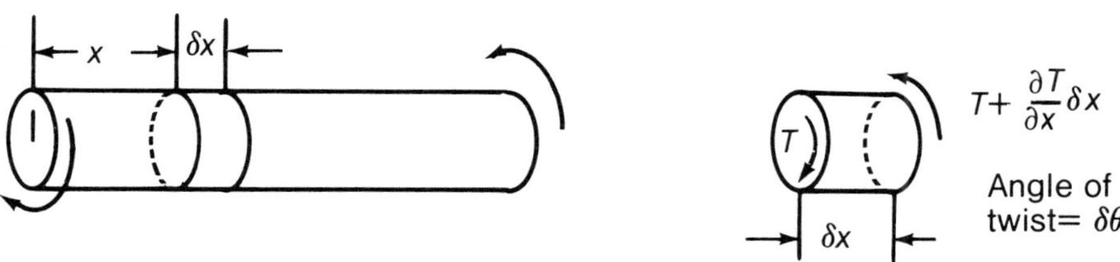

Fig. 6.3(a) Shaft in torsion

Thus, the natural frequencies of the bar in torsional vibration are given by the equation $\omega = (n+\frac{1}{2})\frac{\pi}{L}\sqrt{\left(\frac{G}{\rho}\right)}$, where $n = 0, 1, 2, 3 \ldots$.

EXAMPLE 6.3(i)—illustrative of torsional vibration of a tube

A drill-tube for well-boring is made up of sections 114 mm outside diameter and 97 mm inside diameter made of steel of density 7.85 Mg/m^3 and modulus of rigidity 83 GN/m^2. At a depth of 1500 m the cutting bit has a moment of inertia I_b of 40 kg m^2. Estimate the first and second natural frequencies of torsional vibration of the tube assuming it to be fixed at the upper end.

The boundary conditions are, for the upper end, $x = 0$, $\theta = 0$, and, for the lower end, the torque on the tube is equal to the inertia torque of the cutting bit $\left(I_b \frac{d^2\theta}{dt^2} = I_b\omega^2\theta_{x=l}\right)$. Hence, in the equation for angle of twist at any section x the constant $B = 0$, and the lower boundary condition is such that $I_b\omega^2\theta_{x=l} = T = GJ\left(\frac{\partial\theta}{\partial x}\right)_{x=l}$

Hence

$$\theta_{x=l} = \frac{GJ}{I_b\omega^2}\left(\frac{\partial\theta}{\partial x}\right)_{x=l}$$

i.e.

$$\sin\omega t\left\{A\sin\left(\omega l\sqrt{\frac{\rho}{G}}\right)+0\right\} = \frac{GJ}{I_b\omega^2}\left(\sin\omega t\left\{A\omega\sqrt{\frac{\rho}{G}}\cos\left(\omega l\sqrt{\frac{\rho}{G}}\right)\right)\right.$$

or

$$\tan\left(\omega l\sqrt{\frac{\rho}{G}}\right) = \frac{J\sqrt{(\rho G)}}{I_b\omega}$$

and in the case given, $\tan(0.463\,\mathrm{s}\omega) = \frac{5.0}{\omega\mathrm{s}}$ which is solved graphically as $\tan(0.463\,\phi) = \frac{5.0}{\phi}$ in Fig. 6.3(b), where $\phi = \omega\mathrm{s}$. The points of intersection on the graph give the first natural frequency of the vibrating tube

$$\omega_1 = 2.4\frac{\mathrm{rad}}{\mathrm{s}}\left[\frac{\mathrm{cycle}}{2\pi\,\mathrm{rad}}\right] = 0.382 \text{ cycle/s or } 0.382\text{ Hz}$$

and the second $\omega_2 = 8.1$ rad/s $= 1.29$ Hz.

EXAMPLE 6.3(ii)—illustrative of torsional vibration of tapered shafts

Deduce the wave equation for torsional autonomous vibration in a right circular shaft with variable cross section, and compare it with that governing the longitudinal autonomous vibration of an elastic bar (Example 6.1(i))

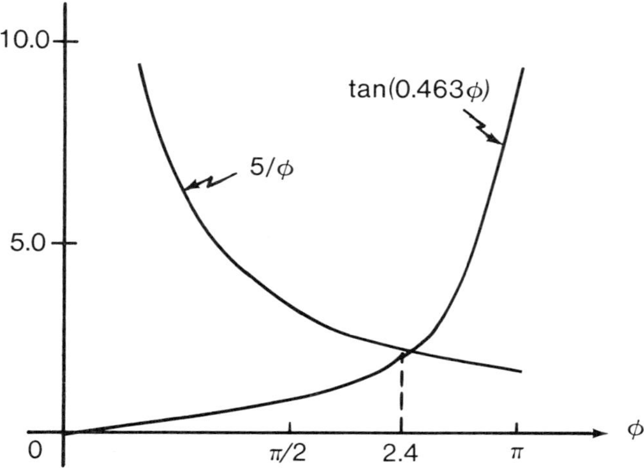

Fig. 6.3(b) Graphical solution of $\tan(0.463\phi) = 5/\phi$

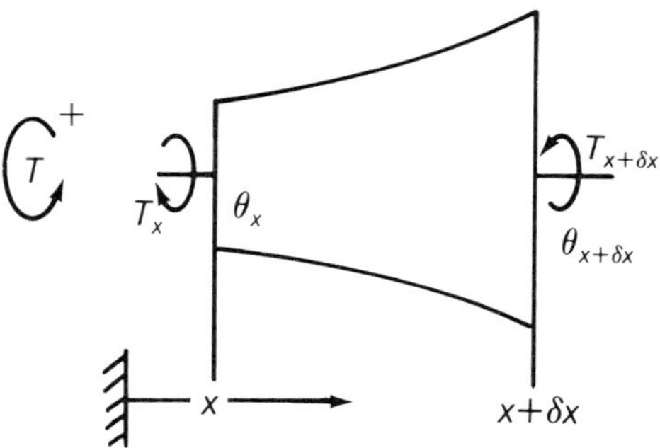

Fig. 6.3(c) Element of length of tapered shaft

Referring to Fig. 6.3(c) and assuming homogeneous (constant ρ), isotropic (constant G) material, and cross-sections remain plane when twisted, then because of the dynamic nature of the system, the local angle of twist, θ, must be a function of location x and time t, i.e. $\theta = f(x, t)$. Hence, by Newton's second law of motion, $T = I\alpha$, applied to the element and using $\dfrac{T}{J} = \dfrac{G\theta}{l} = G\dfrac{\partial\theta}{\partial x}$

$$T_{x+\delta x} - T_x = \rho\,\delta x J_x \frac{\partial^2\theta}{\partial t^2},$$

where, at section $x + \delta x$, $T_{x+\delta x} = T_x + \dfrac{G\partial}{\partial x}\left(J\dfrac{\partial\theta}{\partial x}\right)\delta x$. Hence,

$$\left(J\frac{\partial^2\theta}{\partial x^2} + \frac{\partial J}{\partial x}\frac{\partial\theta}{\partial x}\right) = \frac{\rho}{G}J_x\frac{\partial^2\theta}{\partial t^2}$$

is the wave-equation for torsional autonomous vibrations in a right circular shaft of variable cross section. It is of the same form as that for longitudinal vibration and, as expected, when $\frac{\partial J}{\partial x} = 0$, reduces to the previous equation in this section, namely, $\frac{\partial^2 \theta}{\partial t^2} = \frac{G}{\rho}\frac{\partial^2 \theta}{\partial x^2}$.

6.4 LABORATORY TEST OF A VIBRATING THIN BEAM; AND ILLUSTRATIVE OF THE SIX STAGES IN LABORATORY WORK (SEE EXAMPLE 1.4(ii) FOR THE STAGES OF THEORY)

(I) Object (1) To compare the theoretical profile of a 'free-free' thin steel beam with that obtained by observation. (2) To determine the value of E of the steel.

(II) Apparatus and method of testing A steel beam 750 mm long 50 mm wide and 8 mm deep is supported by two horizontal wires at nodes (found by putting $y = 0$ in the following formula (4)). Vibration is maintained by means of an electric circuit shown diagrammatically in Fig. 6.4(a) and includes a counter for frequency measurement. A series of centre-pop marks 20 mm apart on the side of the beam appear as vertical lines when the beam vibrates and when observed through a telescope show the amplitudes which are measured to obtain the actual profile. Observe and measure the amplitudes on half of the beam and keep another telescope on a point at the end of the beam as a control on the amplitude during readings. Also check the frequency of vibration during reading of amplitudes—say, note the time for 3000 vibrations (e.g. 42.5 s).

Theoretically, on the assumptions that (i) the beam is shallow and without end-thrust, (ii) the usual formulae for bending of beams apply (see Appendix A4) and (iii) the motion is simple harmonic, the equation of motion is

$$\frac{\partial F}{\partial x}\delta x = \rho a \delta x \frac{\partial^2 y}{\partial t^2} \tag{1}$$

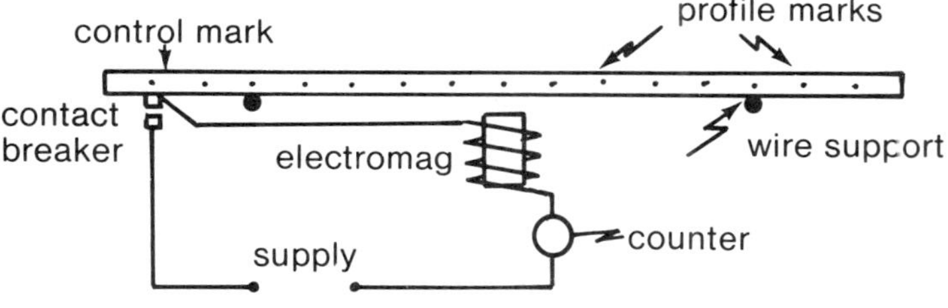

Fig. 6.4(a) Free-free vibrating beam

and for the elastic beam

$$M = EI\frac{\partial^2 y}{\partial x^2} \quad \text{and} \quad F = -\frac{\partial M}{\partial x} \tag{2}$$

therefore

$$\frac{\partial F}{\partial x} = -EI\frac{\partial^4 y}{\partial x^4} \tag{2a}$$

and for harmonic motion

$$\frac{\partial^2 y}{\partial t^2} = -\omega^2 y \tag{3}$$

where $y = u\sin(\omega t + \alpha)$ and $u = \psi(x)$. Hence, from 1, 2a and 3, we deduce that

$$EI\frac{\partial^4 y}{\partial x^4} = \rho a\omega^2 y, \text{ or } \frac{\partial^4 y}{\partial x^4} - \lambda^4 y = 0, \text{ where } \lambda^4 = \frac{\rho a\omega^2}{EI}$$

the solution of which (see Appendix B12(v) is $y = A\cos\lambda x + B\sin\lambda x + C\cosh\lambda x + D\sinh\lambda x$.

For a 'free-free' beam supported at nodes, the end conditions are $x = 0$, $M = 0$, $F = 0$ and $x = l$, $M = 0$, $F = 0$. Hence, using

$$\frac{M}{EI} = \frac{\partial^2 y}{\partial t^2} \quad \text{and} \quad \frac{F}{EI} = -\frac{\partial^3 y}{\partial x^3},$$

we deduce that $C = A$, $D = B$, $\dfrac{A}{B} = -\dfrac{\sinh\lambda l - \sin\lambda l}{\cosh\lambda l - \cos\lambda l}$ and $\cosh\lambda l\cos\lambda l = 1$. From the latter equation we find that $\lambda l = 0$; 4.73; 7.85; 10.99; etc. Hence, for $\lambda l = 4.73$, $\dfrac{A}{B} = -1.018$ and

$$\frac{y}{A} = (\cos\lambda x + \cosh\lambda x) - \left(\frac{\sin\lambda x + \sinh\lambda x}{1.018}\right) \tag{4}$$

The nodes are found by putting $y = 0$ and graphing for the solution $\lambda x = 1.058$. Hence, $\dfrac{\lambda x}{\lambda l} = \dfrac{1.058}{4.73} = 0.224$ which gives $x = 0.224l$ as the distance of the lowest normal mode from the end of the beam—i.e. the position of the supporting wires.

III Observations Taking measurements from mid-span, i.e. $l/2 = 375$ mm or $z = 0$ as in Fig. 6.4(b), then, if at mid-span, $y = 6.6$ divisions in the eyepiece of the microscope, $A = -5.46$. Actual readings in divisions can be taken of the several 'centre-pop' marks, tabulated and plotted where $z = x - 375$ mm.

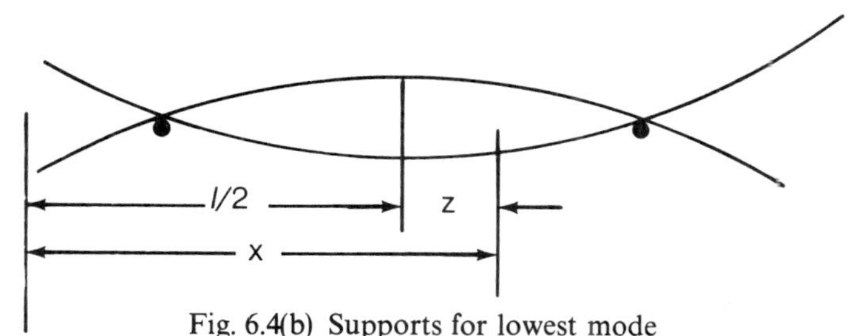

Fig. 6.4(b) Supports for lowest mode

e.g. z/mm	0	20	40	...	360
y/divisions	6.6				

IV Calculations The theoretical profile is obtained from a plot of

$$y = -5.46\{(\cos\lambda x + \cosh\lambda x) - \frac{1}{1.018}(\sin\lambda x + \sinh\lambda x),$$

where

$$\lambda x = \frac{4.73}{750}\left(\frac{x}{\text{mm}}\right).$$

The value of E may be calculated from $\lambda^2 = \omega\sqrt{\frac{\rho a}{EI}}$ or $E = \frac{\omega^2\rho}{k^2\lambda^4}$, where $\omega = 2\pi/\tau = 2\pi f$.

V Graphs The calculated and observed values of y or y/A can be plotted on a common base of x or z, and compared.

VI Conclusions can be drawn as to the accuracy of theoretical predictions.

EXAMPLE 6.4(i)—illustrative of lateral vibration of a flat plate
It is proposed to set up a model to study 'panting' phenomena. The model consists of a flat steel plate 3000 mm × 750 mm × 12 mm stiffened with two steel I beams 3000 mm long welded to the plate. The I beams are 150 mm deep with flanges 75 mm wide and have a mass of 18 kg/m and the relevant second moment of area is 870 cm^4 each having a cross-sectional area of 22.5 cm^2. Estimate the frequency of undamped natural vibration in the 'free-free' mode assuming $\lambda l = 4.73$ and $E = 200$ GN/m^2.

The common centre of gravity, referring to Fig. 6.4(c), is given by $750 \times 12(75 - x + 6) = 4500\,x$ i.e. $x = 54$ mm.

Hence, the second moment of area of the system is

$$I = 2(870\,\text{cm}^4 + 22.5\,\text{cm}^2 \times 5.4^2\,\text{cm}^2) + 75 \times 1.2\,\text{cm}^2(8.1\,\text{cm} - 5.4\,\text{cm})^2$$
$$= 2052\,\text{cm}^4 + 657\,\text{cm}^4 = 2709\,\text{cm}^4$$

The density of the material is $18\,\text{kg/m} \times \frac{10^6}{2250\,\text{m}^2} = 8000\,\text{kg/m}^3$. The natural frequency

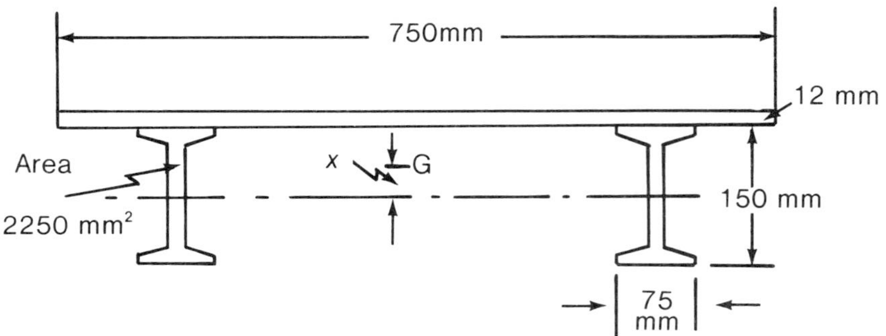

Fig. 6.4(c) Stiffened plate in cross-section

of the system in the lowest mode of 'free free' vibration ($\lambda l = 4.73$) is given by

$$\omega^2 = \lambda^4 \frac{EI}{\rho a} = \left(\frac{4.73}{3\,\text{m}}\right)^4 \times 200\frac{\text{GN}}{\text{m}^2} \times 2709\frac{\text{m}^4}{10^8} \times \frac{\text{m}^3}{8000\,\text{kg}} \times \frac{10^6}{2250\,\text{m}^2}$$

$$= 1.66 \times 10^6 \frac{\text{N}}{\text{kg}\,\text{m}} = 1.66 \times 10^6/\text{s}^2$$

$$\text{Hence, } \omega = 1290/\text{s}\left[\frac{\text{cycle}}{2\pi}\right] = 205 \text{ cycle/s} = 205 \text{ Hz}$$

6.5 SYSTEMS WITH SEVERAL DEGREES OF FREEDOM

In general we may consider a system with n degrees of freedom and its principal modes. Using 'influence coefficients', namely, defining symbol a_{ij} to represent the deflection at a point i caused by a unit load at j (all other loads being zero), then a_{ii} represents the deflection at i of a unit load at i; and a_{ii} is the reciprocal of the stiffness k_i of the system at the point i.

If x_i is the deflection of mass m_i at the point i in a system vibrating in one of its principal modes with frequency ω, the inertia force is $m_i x_i \omega^2$. In particular, at point No 1 the deflection may be written

$$x_1 = a_{11}(m_1 x_1 \omega^2) + a_{12}(m_2 x_2 \omega^2) + a_{13}(m_3 x_3 \omega^2) + \cdots.$$

and similarly for other points 2, 3, ... n of the system which, using the matrix notation (see Appendix B3), can conveniently be written:

$$\begin{Bmatrix} x_1 \\ x_2 \\ x_3 \\ \vdots \\ x_n \end{Bmatrix} = \omega^2 \begin{bmatrix} a_{11} & m_1 & a_{12} & m_2 & a_{13} & m_3 & - & - \\ a_{21} & m_1 & a_{22} & m_2 & a_{23} & m_3 & - & - \\ a_{31} & m_1 & a_{32} & m_2 & a_{33} & m_3 & - & - \\ - & - & - & - & - & - & - & - \\ - & - & - & - & - & - & - & - \\ a_{n1} & m_1 & a_{n2} & m_2 & a_{n3} & m_3 & - & - \end{bmatrix} \begin{Bmatrix} x_1 \\ x_2 \\ x_3 \\ - \\ - \\ x_n \end{Bmatrix}$$

or, in general, $\{x\} = \omega^2[D]\{x\}$ needing a computer. But taking, here, the case of three degrees of freedom—say, three masses and three springs, $n = 3$ and the three-line determinant when expanded yields the frequency equation:

$$(1/\omega^2)^3 - \{a_{11}m_1 + a_{22}m_2 + a_{33}m_3\}(1/\omega^2)^2$$
$$- \{a_{12}m_2a_{21}m_1 + \cdots\}1/\omega^2 - \{\cdots\} = 0$$

which, if the roots are $\dfrac{1}{\omega_1}$, $\dfrac{1}{\omega_2}$ and $\dfrac{1}{\omega_3}$, can be factorized in the form $\left(\dfrac{1}{\omega^2} - \dfrac{1}{\omega_1^2}\right)\left(\dfrac{1}{\omega^2} - \dfrac{1}{\omega_2^2}\right)\left(\dfrac{1}{\omega^2} - \dfrac{1}{\omega_3^2}\right) = 0$ which yields

$$\left(\frac{1}{\omega^2}\right)^3 - \left\{\frac{1}{\omega_1^2} + \frac{1}{\omega_2^2} + \frac{1}{\omega_3^2}\right\}\left(\frac{1}{\omega^2}\right)^2 - \{\ldots\}\frac{1}{\omega^2} - \{\ldots\} = 0$$

as the frequency equation (as in Appendix B3).

Comparing the two equations enables us to write

$$\frac{1}{\omega_1^2} + \frac{1}{\omega_2^2} + \frac{1}{\omega_3^2} = a_{11}m_1 + a_{22}m_2 + a_{33}m_3$$
$$= \frac{m_1}{k_1} + \frac{m_2}{k_2} + \frac{m_3}{k_3}$$
$$= \frac{1}{\omega_{11}^2} + \frac{1}{\omega_{22}^2} + \frac{1}{\omega_{33}^2}$$

where ω_{11}, ω_{22}, ω_{33} are natural frequencies of the system with each mass acting separately in the absence of other masses (see Examples 6.5(ii) and (iii)).

Also, since ω_2 and ω_3 are natural frequencies corresponding to higher modes and larger than the fundamental (ω_1), we may neglect them in an equation for the approximate fundamental frequency. Hence, extending this to any number of degrees of freedom we arrive at Dunkerley's equation for estimating the approximate value, ω_1, of the fundamental frequency of systems with several degrees of freedom, namely:

$$\frac{1}{\omega_1^2} \doteqdot \frac{1}{\omega_{11}^2} + \frac{1}{\omega_{22}^2} + \frac{1}{\omega_{33}^2} + \cdots = a_{11}m_1 + a_{22}m_2 + a_{33}m_3 + \cdots$$

Alternatively, in terms of periodic time:

$$\tau_1^2 = \tau_{11}^2 + \tau_{22}^2 + \tau_{33}^2 + \cdots$$

These latter equations have many useful applications—e.g. in the case of a small structure, say the wing or rudder of an aeroplane, of lowest natural or fundamental frequency ω_{11} and ω_{22} is the natural frequency of an exciter or shaker to be mounted on the structure (see Example 4.1(i)), then the lowest resonant or natural frequency (ω_1) of the structure when the exciter is mounted on it is given by the equation

$$\frac{1}{\omega_1^2} = \frac{1}{\omega_{11}^2} + \frac{1}{\omega_{22}^2} = \frac{1}{\omega_{11}^2} + a_{22}m_2 = \frac{1}{\omega_{11}^2} + \frac{m_2}{k_2}$$

from which ω_{11}, the approximate fundamental frequency of the structure, can be found by eliminating a_{22} as a result of two observations (see Example 6.5(i)).

Thus, Dunkerley's equation enables an estimate of the lowest natural frequency (ω_{11}) of a structure to be made by attaching an eccentric-mass exciter to the structure and measuring the frequency (ω_1) corresponding to the maximum amplitude.

The frequency (ω_1) of the structure plus exciter differs from the natural frequency of the structure itself especially if the mass of the exciter is a substantial part of the total mass vibrating. However, since $\frac{1}{\omega_{11}^2} = \frac{1}{\omega_1^2} - a_{22}m_2$, where m_2 is the mass of the exciter and a_{22} is the influence coefficient of the structure at the point of attachment of the exciter, a_{22} can be eliminated by measuring the two values of ω_1 which correspond to the two maximum amplitudes when two different eccentric-masses are used in the exciter.

EXAMPLE 6.5(i)—illustrative of a structure vibrating

A rotating eccentric-mass of 0.5 kg excites a small structure into resonance at a frequency of 30 Hz. By doubling the mass of the exciter, the resonant frequency was reduced to 24 Hz. Estimate the approximate fundamental frequency of the structure and the stiffness at the point of attachment of the exciter.

Using Dunkerley's equation in the form $\frac{1}{\omega_1^2} = \frac{1}{\omega_{11}^2} + a_{22}m_2$, we have in the first case

$$\omega_1 = 30\,\mathrm{Hz}\left[\frac{2\pi\,\mathrm{rad}}{\mathrm{Hz\,s}}\right] = 60\,\pi/\mathrm{s}$$

and, hence,

$$\left(\frac{\mathrm{s}}{60\,\pi}\right)^2 = \frac{1}{\omega_{11}^2} + a_{22} \times 0.5\,\mathrm{kg}.$$

In the second case, $\left(\frac{\mathrm{s}}{48\pi}\right)^2 = \frac{1}{\omega_{11}^2} + a_{22} \times 1.0\,\mathrm{kg}$. Thus, eliminating a_{22}, we have the primary natural frequency of the structure given by

$$\left(\frac{s}{2\pi}\right)^2\left\{\frac{2}{30^2}-\frac{1}{24^2}\right\}=\frac{1}{\omega_{11}^2}=\left(\frac{1}{2\pi f_{11}}\right)^2 \quad \text{and} \quad f_{11}=45.3\,\text{Hz.}$$

The stiffness of the structure at the point of attachment of the exciter is

$$k_2=\frac{1}{a_{22}}=31.6\,\text{N/mm.}$$

EXAMPLE 6.5(ii)—illustrative of fundamental frequency of a loaded cantilever
Using Dunkerley's equation, estimate the fundamental frequency of a uniform cantilever beam with a mass, m, concentrated at the end and equal to the mass of the beam.

Referring to Fig. 6.5(a) and Example 6.2(i), the frequency of the beam alone is given by $\omega_{11}^2=(3.54)^2\left(\frac{EI}{ml^3}\right)$ and for the mass, m, alone $\omega_{22}^2=3.0\left(\frac{EI}{ml^3}\right)$ as in Example 2.3 (iv).
Hence, $\frac{1}{\omega_1^2}\doteqdot\frac{1}{\omega_{11}^2}+\frac{1}{\omega_{22}^2}$ and the fundamental natural frequency of the system is given by.

$$\omega_1^2=\frac{\omega_{11}^2\times\omega_{22}^2}{\omega_{11}^2+\omega_{22}^2}=\left\{\frac{(3.54)^2(3.0)}{(3.54)^2+(3.0)}\right\}\frac{EI}{ml^3}=2.45\left(\frac{EI}{ml^3}\right)$$

The result compares favourably with that obtained using Rayleigh's method (see Example 2.5(iii)).

EXAMPLE 6.5(iii)—illustrative of vibration of machines and their mountings
Two beams each of which weighs 235 N/m, are supported on rigid supports having a span of 3.6 m. They carry a motor-generator comprising two machines, each weighing 5 kN. The centres of gravity of the machines are located symmetrically 0.6 m left and right of mid-span. When the first machine is lowered on to the composite beam, the latter deflects 1.6 mm immediately below the centre of gravity of that machine. Estimate the approximate frequency of natural vibration of the complete set.

The deflection of a simply-supported beam at the point where it carries a load W located at distances a and b from the supports, respectively, is given by

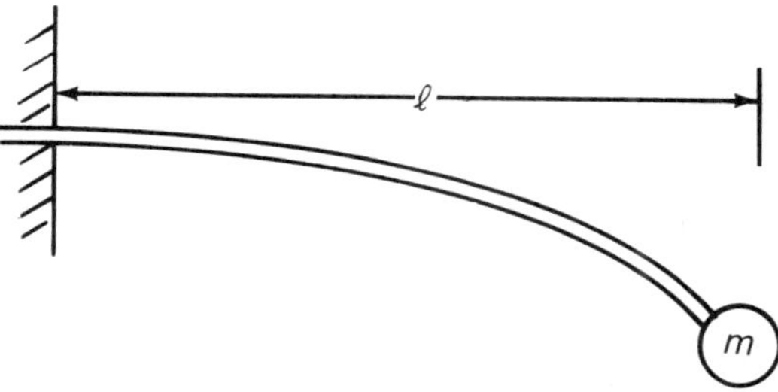

Fig. 6.5(a) Vibrating mass on a cantilever beam

$\delta = \dfrac{Wa^2b^2}{3EI(a+b)}$. Hence, for the composite beam in question,

$$EI = \frac{5\text{ kN} \times 1.44\text{ m}^2 \times 4 \times 1.44\text{ m}^2}{3 \times 3.6\text{ m} \times 1.6\text{ mm}} = 2.4\text{ MNm}^2.$$

The lowest natural frequency of the composite unloaded beam whose weight per unit length is $w = \left(\dfrac{\rho alg}{l}\right) = 470\text{ N/m}\left(\text{or mass per unit length } \rho a = \dfrac{470\text{ kg}}{9.81\text{ m}} = 48\text{ kg/m}\right)$ is $\omega_{11} = (\lambda l)^2 \sqrt{\left(\dfrac{EI}{\rho al^4}\right)}$, where $\lambda l = \pi$ for a simply-supported beam,

$$= \frac{\pi^2}{(3.6\text{ m})^2} \sqrt{\left(\frac{2.4\text{ MN m}^2}{48\text{ kg/m}}\left[\frac{\text{kg m}}{\text{N s}^2}\right]\right)} = \frac{0.762}{\text{m}^2} \sqrt{\left(\frac{10^6\text{ m}^4}{20\text{ s}^2}\right)} = 170/\text{s}.$$

The natural frequency of the load where the elastic stiffness $k = W/\delta$ at the points where each of the two loads are applied is given by $\omega_{22}^2 = k/m = g/d = \omega_{33}^2$ i.e.

$$\frac{1}{\omega_{22}^2} = \frac{1}{\omega_{33}^2} = \frac{d}{g} = \frac{1.6\text{ mm}}{9807\text{ mm/s}^2} = \frac{\text{s}^2}{6130}.$$

Hence, using Dunkerley's equation, the lower frequency (ω_1) of the complete set is given by

$$\frac{1}{\omega_1^2} = \frac{1}{\omega_{11}^2} + \frac{1}{\omega_{22}^2} + \frac{1}{\omega_{33}^2} = \frac{\text{s}^2}{10^4}\left(\frac{1}{2.89} + \frac{2}{0.613}\right) = \frac{\text{s}^2}{10^4} \times 3.6$$

i.e. $\omega_1 = 52.6/\text{s}$ or $\dfrac{52.6}{\text{s}}\left[\dfrac{\text{cycle}}{2\pi}\right] = 8.37$ cycles/s or 8.37 Hz.

For suggestions on further reading and study, see Appendix D, (6), (7) and (8).

EXERCISES ON CHAPTER 6

1. Determine the natural frequencies of a uniform beam of length l, clamped at one end and pinned at the other.
2. A uniform beam of length l and mass M_b is clamped at one end, and carries a concentrated mass M_0 at the other end. Find the frequency formula.
3. Determine the expressions for the natural frequencies and the node position for the lowest frequency of a free-free bar of length l in lateral vibration.
4. A concrete test-beam 50 mm × 50 mm × 300 mm supported at two points $0.224l$ from the ends, was found to resonate at 1590 cycles/s. If the density of concrete is 2450 kg/m^3, determine the modulus of elasticity, assuming the analysis for slender beams applies.
5. It is proposed to set up a model of a deck plate of a ship with two I beams welded to it, and to set this in natural vibration in its primary free-free mode to study fatigue. The plate will measure 250 cm × 75 cm × 13 mm, and each of the I beams weighing 180 N/m will measure 250 cm × 15 cm × 7.5 cm each having a relevant second moment of area 876 cm^4. Neglecting the I of the plate, estimate the frequency of

undamped natural vibration of the rig assuming that the value of λl appropriate to the free-free primary mode is 4.73. $E = 200\ \text{GN/m}^2$.

6. The natural frequency of the wing of an aeroplane in torsion is 25 cycles/s. A missile of mass 450 kg is hung in a position where the torsional stiffness of the wing is 7×10^6 Nm/rad, and moment of inertia of the suspended missile about the torsional axis is 200 kg m^2. Using Dunkerley's equation, estimate the approximate torsional frequency of the loaded wing.

7. State the meaning of Dunkerley's equation $\dfrac{1}{\omega_1^2} = \dfrac{1}{\omega_{11}^2} + \dfrac{1}{\omega_{22}^2} + \dfrac{1}{\omega_{33}^2} + \cdots$, or $\tau^2 = \tau_1^2 + \tau_2^2 + \tau_3^2 + \cdots \doteqdot m_1/k_1 + m_2/k_2 + m_3/k_3 + \cdots \doteqdot a_{11}m_1 + a_{22}m_2 + \cdots$.

   Hence, find the approximate frequency of a beam of length l simply-supported having two concentrated loads each of mass m symmetrically spaced at a distance $l/3$ apart.

   If the beam is steel of circular cross-section 50 mm diameter and 3 m long and each mass is 180 kg, calculate the fundamental, frequency assuming $E = 207\ \text{GN/m}^2$ and density of beam 7900 kg/m^3, (i) neglecting the mass of the beam, and (ii) taking the mass of the beam into account.

8. A beam vibrated by an eccentric-mass shaker of mass 5 kg at mid-span was found to resonate at 440 Hz.

   With an additional mass of 5 kg the resonant frequency was lowered to 400 Hz. Estimate the natural frequency of the beam, and the stiffness of the beam at mid-span based on the fact that resonant frequency is due to the total mass of the beam and shaker.

Appendix A

Mechanics—proofs of formulae used in the text

A.1 CHANGES IN VELOCITY, MOMENTUM AND ENERGY

(a) Acceleration of a body moving in a two-dimensional curve

Acceleration is produced by change in direction as well as by change in speed. Referring to Fig. A.1(a) the mean acceleration of a body moving in a curve during the interval t to $t + \delta t$ is $\dfrac{\text{AB}}{\delta t}$ in direction of length AB. This can be resolved into the two components (i) tangential at O $= \underset{\delta t \to 0}{\text{Lt}} \dfrac{\text{AC}}{\delta t} = a$ in direction AC, and (ii) centripetal at O $= \underset{\delta t \to 0}{\text{Lt}} \dfrac{\text{CB}}{\delta t}$ in direction CB, and, since $\text{CB} \doteqdot (\text{OB})\delta\psi \doteqdot \dfrac{(\text{OB})}{r}\dfrac{\delta x}{\delta t}$, then the centripetal acceleration at O is

$$\underset{\delta t \to 0}{\text{Lt}} \frac{(v + \delta v)}{r} = \frac{v^2}{r} = r\omega^2,$$

where ω is rad/unit time (also see B.8),

(b) Principles concerning momentum and energy

Principles concerning momentum and energy are both only one step removed from Newton's second law of motion $F = ma$, where F is the force causing an acceleration a on a particle of mass m. Thus,

$$F = ma = m\frac{\mathrm{d}v}{\mathrm{d}t} = m\frac{\mathrm{d}v}{\mathrm{d}s}\cdot\frac{\mathrm{d}s}{\mathrm{d}t} = mv\frac{\mathrm{d}v}{\mathrm{d}s}$$

Hence (i) $\int_1^2 F\,\mathrm{d}t = \int_1^2 m\,\mathrm{d}v$ or $\int_1^2 F\,\mathrm{d}t = mv_2 - mv_1$, i.e. the change in momentum in a given time is equal to the time integral of the applied force over this period, i.e. to the impulse, and (ii) $\int_1^2 F\,\mathrm{d}s = \int_1^2 mv\,\mathrm{d}v = \frac{1}{2}mv_2^2 - \frac{1}{2}mv_1^2$, i.e. the change in

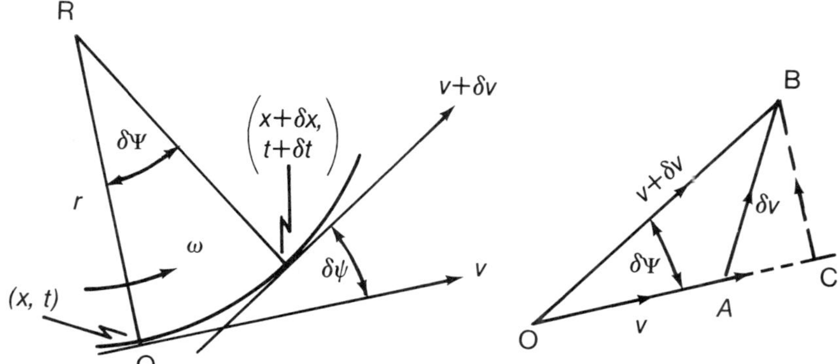

Fig. A.1(a) Acceleration in plane of motion

kinetic energy in a given distance is equal to the space integral of the applied force over this distance, or the work done.

A.2 TWO-DIMENSIONAL MOTION ABOUT A FIXED AXIS

(as used in Section 1.4 and others)

Referring to Fig. A.2(a), the solid body of mass M whose centre of mass is at G moves about a fixed horizontal axis through O.

(**a**) The component forces δX and δY act on the particle of mass δm. Hence, $\sum\delta m\ddot{x} = \sum\delta X$, i.e. $M\ddot{x}_G = X$ since $\sum\delta mx = Mx_G$ by definition of G. Similarly $M\ddot{y}_G = Y$. These equations show that the motion of the body of mass M is the same as that of a mass M concentrated at G acted on by component forces X and Y.

(**b**) The rate of change of angular momentum of the body about O is $\frac{\mathrm{d}}{\mathrm{d}t}\sum\delta m(\dot{y}x - \dot{x}y) = \sum\delta m(\ddot{y}x - \ddot{x}y)$ which, using the results of (a), reduces to

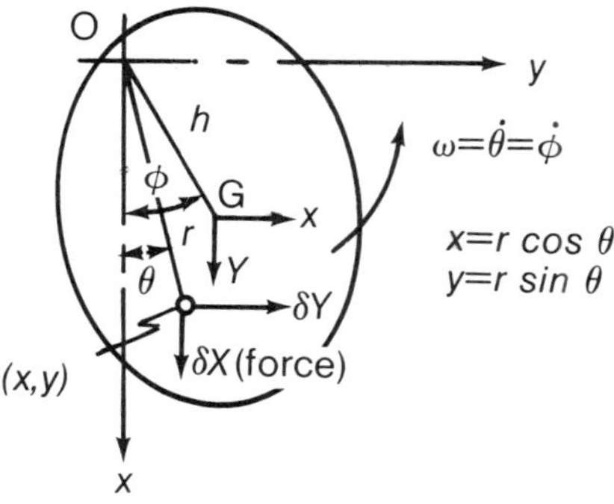

Fig. A.2(a) Motion about a fixed axis

$\sum(\delta Yx - \delta Xy) = Yx_G - Xy_G$ which is the moment (T_o) of the external forces about O. Also, since

$$\sum\delta m(\dot{y}x - \dot{x}y) = \sum\delta m(r^2\dot{\theta}\cos^2\theta + r^2\dot{\theta}\sin^2\theta)$$
$$= \dot{\theta}\sum\delta mr^2 = I_o\dot{\theta},$$

$\frac{\mathrm{d}}{\mathrm{d}t}(I_o\dot{\theta}) = T_o = I_o\ddot{\theta}$ or $I_o\alpha$, where α is the angular acceleration.

(c) The rate of increase of kinetic energy of the body is $\frac{\mathrm{d}}{\mathrm{d}t}\sum\frac{1}{2}\delta m(r\dot{\theta})^2 =$ $\frac{\mathrm{d}}{\mathrm{d}t}(\frac{1}{2}I_O\dot{\theta}^2) = \dot{\theta}T_O = \dot{\theta}(Yx_G - Xy_G)$ using the results of (b). Alternatively, since $x_G = h\cos\phi$ and $y_G = h\sin\phi$, and $\dot{\theta} = \dot{\phi}$ the latter expression can be written $\dot{\theta}h\,(Y\cos\phi - X\sin\phi) = Y\dot{y}_G + X\dot{x}_G =$ the rate at which external forces do work ($\dot{\theta}T_O$). Thus, the rate of increase of kinetic energy of the body is $T_O\omega$.

A.3 RIGID BODY MOVING FREELY IN TWO DIMENSIONS

(as used in Section 1.4, etc)

So far as the externally-applied system of forces is concerned, we shall prove that the only difference between particle dynamics (see Appendix A.1) and two-dimensional rigid dynamics is the addition, in the latter case, of the couple or torque $T_G = I_G\ddot{\theta}$ to $F = Ma_G$.

(a) Referring to Fig. A.3(a), the component forces δX and δY act on particle δm hence, $\sum\delta X = \sum\delta m\ddot{x}$ or $X = M\ddot{x}_G$ and $Y = My_G$.

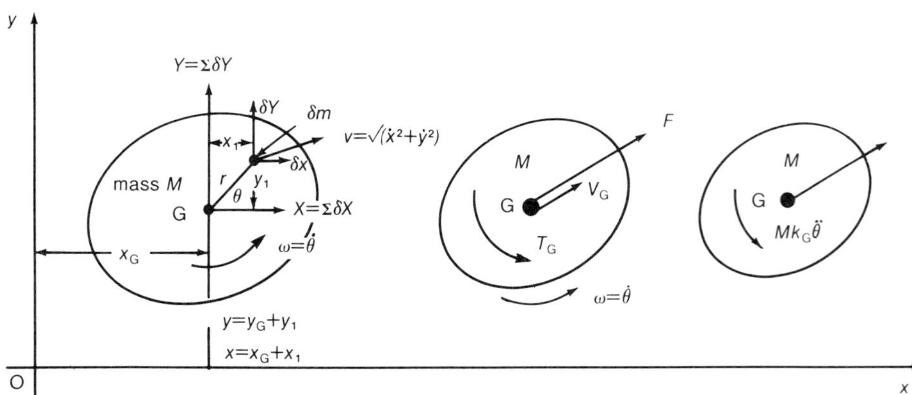

Fig. A.3(a) Free movement in two dimensions

(b) The rate of change of angular momentum about any point O is $\frac{\mathrm{d}}{\mathrm{d}t}\sum\delta m(\dot{y}x - \dot{x}y) = \sum(\delta Y x - \delta X y) = Yx_G - Xy_G$ which is the moment of the external forces about that point. (This equation does not reduce to $T_O = I_O\ddot{\theta}$ unless the motion is restricted to that about a fixed axis as in Section A.2). The equation can, however, be recast with G (the centres of mass) as the reference point to give $T_G = I_G\ddot{\theta}$ for a body moving freely in two dimensions. The rate of change of angular momentum about any reference point O may be written $\sum\delta m\{(\ddot{y}_G + \ddot{y}_1)(x_G + x_1) - (\ddot{x}_G + \ddot{x}_1)(y_G + y_1)\} = \sum\{\delta Y(x_G + x_1) - \delta X(y_G + y_1)\}$ and (since by definition of point G, $\sum\delta m x_1 = 0 = \sum\delta m y_1$ and $\sum\delta m\ddot{x}_1 = 0 = \sum\delta m\ddot{y}_1$) alternatively as

$$\begin{aligned} &M(\ddot{y}_G x_G - \ddot{x}_G y_G) + \sum\delta m(\ddot{y}_1 x_1 - \ddot{x}_1 y_1) \\ &\quad = \sum\delta Y x_G - \sum\delta X y_G + \sum(\delta Y x_1 - \delta X y_1). \end{aligned}$$

Hence, since $M\ddot{y}_G x_G = \sum\delta Y x_G$ then $\sum\delta m(\ddot{y}_1 x_1 - \ddot{x}_1 y_1) = \sum(\delta Y x_1 - \delta X y_1)$ which is T_G, the moment of the external forces about G. Also, since $x_1 = r\cos\theta$ and $y_1 = r\sin\theta$, then

$$\ddot{x}_1 = -r(\ddot{\theta}\sin\theta + \dot{\theta}^2\cos\theta) \quad \text{and} \quad \ddot{y}_1 = r(\ddot{\theta}\cos\theta - \dot{\theta}^2\sin\theta).$$

Hence, $\sum\delta m\,(\ddot{y}_1 x_1 - \ddot{x}_1 y_1) = \sum\delta m r^2\ddot{\theta}$ or $\boldsymbol{T_G = I_G\ddot{\theta}}$.

We may also note that the angular momentum about any point, namely, $\sum\delta m(\dot{y}x - \dot{x}y)$ may be expressed with reference to the centre of mass, G, as

$$\begin{aligned} &\sum\delta m\{(\dot{y}_G + \dot{y}_1)(x_G + x_1) - (\dot{x}_G + \dot{x}_1)(y_G + y_1)\} \\ &\quad = \sum\delta m(\dot{y}_G x_G - \dot{x}_G y_G) + \sum\delta m(\dot{y}_1 x_1 - \dot{x}_1 y_1) \\ &\quad = M(\dot{y}_G x_G - \dot{x}_G y_G) + \dot{\theta}\sum\delta m r^2 \\ &\quad = M(\dot{y}_G x_G - \dot{x}_G y_G) + I_G\omega \end{aligned}$$

i.e. the angular momentum of a free body moving in two dimensions about any point O is the sum of the moment of momentum of M at G about O and the angular momentum of the body about G.

(c) The rate of increase of total kinetic energy of the body is

$$\begin{aligned} \frac{\mathrm{d}}{\mathrm{d}t}\sum\tfrac{1}{2}\delta m v^2 &= \frac{\mathrm{d}}{\mathrm{d}t}\left\{\sum\frac{\delta m}{2}(\dot{x}^2 + \dot{y}^2)\right\} \\ &= \sum\delta m(\dot{x}\ddot{x} + \dot{y}\ddot{y}) = \sum(\delta X\dot{x} + \delta Y\dot{y}) \\ &= \text{the rate at which external forces do work on the body.} \end{aligned}$$

We may note that the kinetic energy of a rigid body moving freely in two dimensions is

$$\sum\tfrac{1}{2}\delta m\{(\dot{x}_G + \dot{x}_1)^2 + (\dot{y}_G + \dot{y}_1)^2\} = \sum\tfrac{1}{2}\delta m(\dot{x}_G^2 + \dot{y}_G^2)$$
$$+ \sum\tfrac{1}{2}\delta m(\dot{x}_1^2 + \dot{y}_1^2) = \sum\tfrac{1}{2}\delta m(v_G^2) + \sum\tfrac{1}{2}\delta m r^2\dot{\theta}^2$$

i.e. the total kinetic energy is $\frac{1}{2}Mv_G^2 + \frac{1}{2}I_G\omega^2$. Also, if δF is the resultant of the two component forces δX and δY which can always be imagined replaced by an equal force through any selected point together with a couple, the external forces acting on the body of Fig. A.3(a) can be regarded as $\sum\delta X$ and $\sum\delta Y$ acting at G together with a couple $\sum\delta C = C$ producing rotation. Thus, the rate of working of the external forces

$$\begin{aligned}
&= \dot{x}_G\sum\delta X + \dot{y}_G\sum\delta Y + \dot{\theta}\sum\delta C \\
&= \dot{x}_G\sum\delta m\ddot{x} + \dot{y}_G\sum\delta m\ddot{y} + \dot{\theta}C \\
&= M(\dot{x}_G\ddot{x}_G + \dot{y}_G\ddot{y}_G) + I_G\dot{\theta}\ddot{\theta} \\
&= \frac{\mathrm{d}}{\mathrm{d}t}\{\tfrac{1}{2}M(\dot{x}_G^2 + \dot{y}_G^2) + \tfrac{1}{2}I_G\dot{\theta}^2\}
\end{aligned}$$

i.e. $Fv_G + T_G\omega = \dfrac{\mathrm{d}}{\mathrm{d}t}\{\tfrac{1}{2}Mv_G^2 + \tfrac{1}{2}I_G\omega^2\}$, where F is the resultant force of components X and Y acting at G which moves with linear velocity v_G, and T_G is the couple about G which causes an angular velocity ω.

Hence, we conclude that the rate of increase of total kinetic energy (translational and rotational) of a rigid body moving freely in two dimensions is equal to the rate at which external forces do work on the body.

A.4 CURVATURE AND BENDING OF BEAMS (as used in Chapters 2 and 6)

Referring to Fig. A.4(a) we can state that $\dfrac{\mathrm{d}x}{\mathrm{d}s} = \cos\psi = \dfrac{1}{\sec\psi}$,

$$\frac{\mathrm{d}y}{\mathrm{d}x} = \tan\psi \text{ and } \frac{\mathrm{d}^2y}{\mathrm{d}x^2} = \sec^2\psi\frac{\mathrm{d}\psi}{\mathrm{d}x} \text{ and } \sec^2\psi = 1 + \left(\frac{\mathrm{d}y}{\mathrm{d}x}\right)^2$$

The curvature at point P is $\dfrac{\mathrm{d}\psi}{\mathrm{d}s} = \dfrac{1}{R}$ by definition.

i.e.

$$\frac{1}{R} = \frac{\mathrm{d}\psi}{\mathrm{d}x}\frac{\mathrm{d}x}{\mathrm{d}s} = \frac{1}{\sec^3\psi}\frac{\mathrm{d}^2y}{\mathrm{d}x^2} = \frac{\pm\dfrac{\mathrm{d}^2y}{\mathrm{d}x^2}}{\left\{1 + \left(\dfrac{\mathrm{d}y}{\mathrm{d}x}\right)^2\right\}^{3/2}}$$

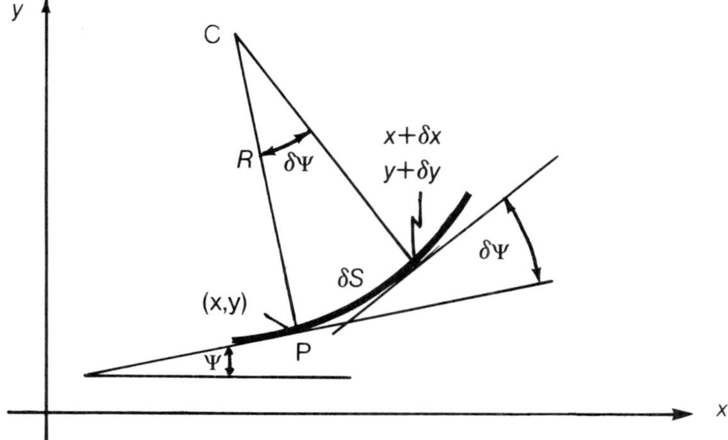

Fig. A.4(a) Curvature

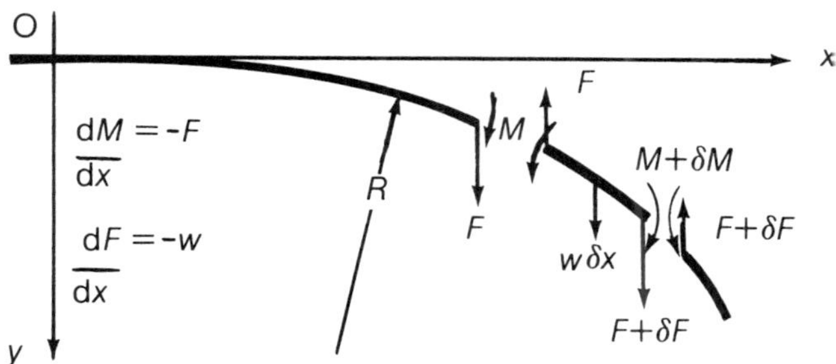

Fig. A4(b) Bending of beams

Hence, if $\left(\frac{\mathrm{d}y}{\mathrm{d}x}\right)$ is relatively small, as in the case of beams, $\frac{1}{R} \doteqdot \pm \frac{\mathrm{d}^2 y}{\mathrm{d}x^2}$ and, referring to Fig. A.4(b) in which the bending moment, M, causes a positive $\frac{\mathrm{d}^2 y}{\mathrm{d}x^2}$, simple beam theory, in which I represents the second moment of area of cross-section, enables us to write

$$\frac{\mathbf{1}}{\boldsymbol{R}} = \frac{\boldsymbol{M}}{\boldsymbol{EI}} \doteqdot \frac{\mathbf{d}^2 \boldsymbol{y}}{\mathbf{d}\boldsymbol{x}^2} \quad \text{and}$$

$$\boldsymbol{EI}\frac{\mathbf{d}^4 \boldsymbol{y}}{\mathbf{d}\boldsymbol{x}^4} = \mathbf{w} = \text{load intensity per unit length.}$$

These results are used in Sections 6.2 and 6.4.

A.5 STRAIN ENERGY OF A BENT BEAM

(as used in Section 2.4 and others)

Magnifying the element of Fig. A.4(b) we get Fig. A.5(a) and deduce that the strain energy (or resilience) of the strip with it under stress f at distance z from the neutral axis is $\frac{1}{2}f^2 x$ volume of strip. Hence, the resilience of the block element of length δx is

$$\frac{1}{2E}\sum f^6 b\delta z\delta x = \frac{1}{2}\left(\frac{M}{I}\right)^2\frac{\delta x}{E}\sum bz^2\delta z = \frac{M^2\delta x}{2EI} = \frac{EI}{2R^2}\delta x = \frac{1}{2}M\delta\theta$$

and the strain energy of the bent beam of length L is $\dfrac{EI}{2}\displaystyle\int_0^L\left(\frac{\mathrm{d}^2 y}{\mathrm{d}x^2}\right)\mathrm{d}x.$

A.6 STRAIN ENERGY OF A SHAFT IN TORSION

Referring to Fig. A.6(a), the work done in twisting the shaft by $\delta\theta$ is $\delta W = FB\delta\theta = T\delta\theta$. Hence, the strain energy stored in a shaft of length L with an angle of twist θ caused by a torque T is $\int_0^\theta T\mathrm{d}\theta = \frac{1}{2}T\theta = \frac{1}{2}\frac{GJ}{I}\theta^2$.

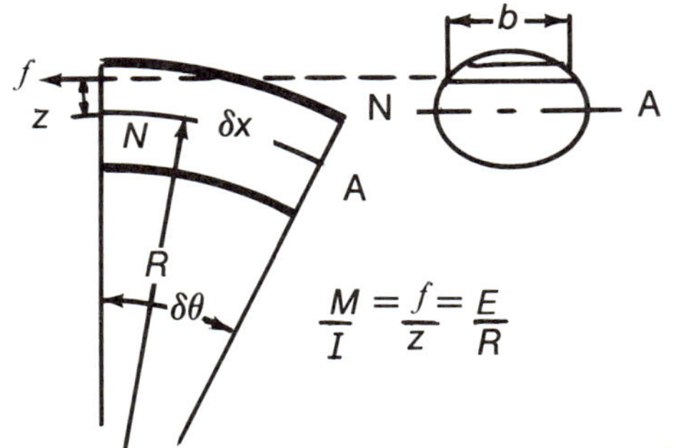

Fig. A.5(a) Bent beam

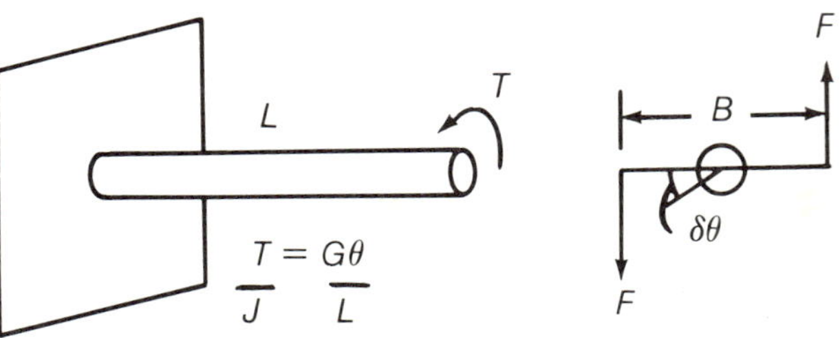

Fig. A.6(a) Shaft in torsion

A.7 MINIMUM STRAIN ENERGY PRINCIPLE

A general principle of mechanics is that a position of equilibrium is one of minimum potential energy—e.g. in the case of an elastic system, of minimum strain energy, as can be verified in the simple case of a stretched spring illustrated in Fig. A.7(a).

The potential energy, V, of the stationary system (due to the strain energy of the stretched spring and the position of the mass relative to the unstretched level) is

$$V = \tfrac{1}{2}kx^2 + W(-x)$$

Hence, $\dfrac{dV}{dx} = kx - W$ which is zero when $x = \dfrac{W}{k}$; and V is a minimum when $W = kx = mg$—i.e. when the system is in the static-equilibrium position.

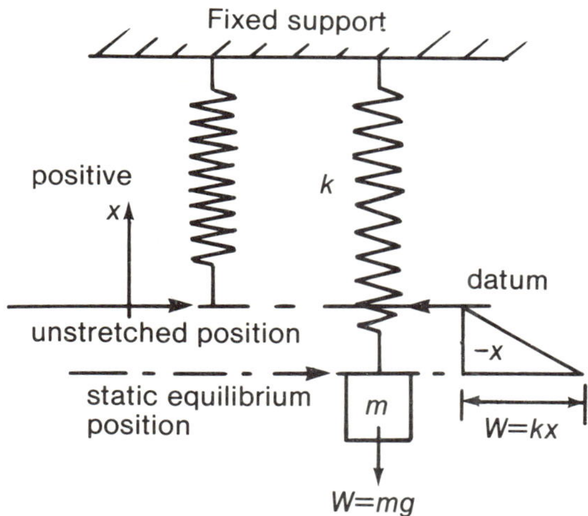

Fig. A.7(a) Energy in a stretched spring

A.8 STIFFNESS OF BEAMS, SHAFTS AND SPRINGS

Stiffness is related to type of material and size. A simply-supported beam of span L and second moment of area I and of material whose modulus of elasticity is E, deflected an amount δ at mid-span due to an applied force F there has a deflection stiffness $k_d = \dfrac{F}{\delta} = \dfrac{48EI}{L^3}$. Similarly, for a circular shaft of length L in torsion, $\dfrac{T}{J} = \dfrac{G\theta}{L}$; hence, the torsional stiffness $k_t = \dfrac{T}{\theta} = \dfrac{GJ}{L}$, where G is the modulus of rigidity of the material, J is the second polar moment of area of the cross-section, and θ is the angle of twist in radians caused by a torque of magnitude T. In the case of helical springs capable of large deflections within a safe stress in the wire of diameter d coiled with diameter D into n turns as shown in Fig. A.8(a), the maximum stress is

$$\frac{Td}{J2} = \frac{\left(\frac{FD}{2}\right)}{\frac{\pi}{32}d^4}\left(\frac{d}{2}\right) = \frac{8FD}{\pi d^3} = \frac{G\theta}{\pi Dn}\left(\frac{d}{2}\right)$$

Also,

$$\theta = \frac{TL}{GJ} = \frac{FD}{G2} \times \frac{\pi Dn}{\frac{\pi}{32}d^4} = \frac{16FD^2n}{Gd^4}$$

hence, deflection

$$\delta = \theta\frac{D}{2} = \frac{8FD^3n}{Gd^4}$$

and the spring stiffness

$$k = \frac{F}{\delta} = \frac{Gd^4}{8D^3n}$$

Fig. A.8(a) Helical spring

Appendix B

Mathematics—operations and solutions used in the text

B.1 DIFFERENT ALGEBRAS AND CONVENTIONS

It is necessary to be aware that, in addition to representing pure numbers (i.e. numerics) and physical quantities (i.e. magnitudes specified by numbers and their units), often the same symbols (e.g. a, A,... t, T,... z, Z: α, A,... ω, Ω) are used (because there aren't enough of them to do otherwise) to represent vector quantities (i.e. magnitude *and* direction), and dimensions only (i.e. their kind or quality as distinct from magnitude).

When there is a possibility of error or confusion by, say, mixed use in analysis and formulae, conventional distinguishing marks can be affixed to the symbols, e.g. $|x|$—the vertical lines signifying the numerical value (in specified units) of a physical quantity x whose magnitude is denoted by a number multiplied by the particular unit used to obtain the number, e.g. if $x = |x|\text{m} = 5\text{m}$, then $|x| = 5$. A bar above a symbol signifies a vector quantity, e.g. a velocity may be denoted as $\bar{v} = 5\text{m/s}$ vertically upwards or in any specified direction; and the dimensions of a linear velocity may be signified as $[v] = [\text{LT}^{-1}]$. Similarly, an angular velocity $\bar{\omega} = 5$ rad/s is a vector quantity having a direction—clockwise or anticlockwise, and dimensional formula $[\omega] = [\text{T}^{-1}]$, since radian is dimensionless.

Thus, four different algebras using the same letter-symbols (e.g. x, v, ...) may be distinguished as:

($|x|$) **numerical algebra,** e.g. $\varepsilon^x = 1 + x + \dfrac{x^2}{2} + \dfrac{x^3}{6} + \cdots$ and $\sin\theta = \theta - \dfrac{\theta^3}{6} + \dfrac{\theta^5}{120} - \cdots$ are numerical series, where x and θ are numbers.

(x) **physical (scalar) algebra,** e.g. $v = \omega r = 5$ m/s, where ω is 'radian-restricted' in that it represents the number of radians per unit time; and $e = |e|\text{Nm} = 3\text{J}$, where $|e| = 3$ and, in this case, Nm is energy (J) not torque, i.e. is the scalar (not the vector) product of force and distance vectors. Thus, many formulae in scalar algebra are resolved vector formulae and relate to magnitudes only as expressed by numbers × their units.

($\bar{x}$) **vector algebra,** e.g. $\bar{v} = 5$ m/s at 48.6° to the abscissa or $\bar{v} = 3\bar{\imath} + 4\bar{\jmath}$, where $\bar{\imath}$ and $\bar{\jmath}$ are unit vectors (each of magnitude 1 m/s) along Cartesian co-ordinates; and $\bar{F} = 3\bar{\imath} + 4\bar{\jmath} + 5\bar{k}$ represents a force of magnitude $(\sqrt{}50)$ N in the corresponding spatial direction.

$[x]$ **dimensional algebra,** e.g. the dimensional equation for dynamic viscosity is $[\eta] = [\mathrm{ML}^{-1}\mathrm{T}^{-1}]$, and for torque is $[T] = [\mathrm{ML}^2\mathrm{T}^{-2}]$ which, although a vector quantity, has the same dimensions as the non-vector (or scalar) product denoting energy. Thus, having the same dimensions does not, necessarily, imply the same kind of physical quantity, as is the case vice versa.

B.2 ROOTS OF QUADRATIC EQUATIONS

$ax^2 + bx + c = 0$ are $x_{1,2} = -\dfrac{b \pm \sqrt{}(b^2 - 4ac)}{2a}$ provided, of course, that $a \neq 0$, (as in Examples 1.5 (ii), 2.6 (ii), etc.)

B.3 SIMULTANEOUS EQUATIONS, DETERMINANTS AND MATRICES.

$$a_1x + b_1y = c_1$$

$$a_2x + b_2y = c_6$$

Yield a solution $x = \begin{vmatrix} c_1 & b_1 \\ c_2 & b_2 \end{vmatrix} \div \begin{vmatrix} a_1 & b_1 \\ a_2 & b_2 \end{vmatrix}$, where the typical determinant $\begin{vmatrix} a_1 & b_1 \\ a_2 & b_2 \end{vmatrix} = a_1b_2 - a_2b_1$ by diagonal multiplication. In general, a new operation in mathematics may suggest itself from practical considerations, or because of advantages within mathematics itself, e.g. matrix notation enables ideas to be generalized within mathematics, and also has important applications outside.

In principle, a matrix is a rectangular array of elements in rows and columns; e.g. a matrix of the order $m \times n$ has m rows and n columns, and each element (or entry) a_{ij} belongs to a set of the same kind (numerical, physical or vector quantities) where suffix i refers to the particular row and j to the particular column. Thus, a $(m \times n)$ matrix $A = (a_{ij})$ means

$$A = \begin{bmatrix} a_{11} & a_{12} \cdots a_{1n} \\ a_{21} & a_{22} \cdots a_{2n} \\ \vdots & \\ a_{m1} & a_{m2} \cdots a_{mn} \end{bmatrix}$$

matrix addition, denoted by ⊞ implies that the matrices must, necessarily, be of the same order e.g. $A \boxplus B = C$ in which $a_{ij} + b_{ij} = c_{ij}$. Matrix multiplication, denoted by □ is the inner product of a row matrix with a column matrix as e.g. defined by

$$(a_{11}a_{12}a_{13})\begin{pmatrix} b_{11} \\ b_{21} \\ b_{31} \end{pmatrix} = a_{11}b_{11} + a_{12}b_{21} + a_{13}b_{31} = \sum_{k=1}^{3} a_{1k}b_{k1}$$

—the row and columns, necessarily, having the same number of elements.

This may seen complicated but is justified by applications especially in the field of linear equations. Thus, $A \square B = C = [c_{ij}]$, i.e. the inner product of the ith row of A and the jth column of B as expressed symbolically by $c_{ij} = \sum_{k=1}^{p} a_{ik} b_{kj}$, where p is the number of columns in A and rows in B. Thus, referring to the types in Examples 2.6 (i) and to Figs. 2.6 (c), 5.1 (b) namely, a (2×2) square matrix $A = \begin{bmatrix} a_{11}a_{12} \\ a_{21}a_{22} \end{bmatrix}$ and a (2×2) column matrix $B = \begin{Bmatrix} b_{11} \\ b_{21} \end{Bmatrix}$, their product $A \square B = \begin{bmatrix} a_{11}b_{11} + a_{12}b_{21} \\ a_{21}b_{11} + a_{22}b_{21} \end{bmatrix}$.

Simultaneous equations can always be put in the matrix form, e.g. the two equations with two unknowns

$\begin{cases} ax + by = p, \text{ where } a \text{ and/or } b \neq 0 \\ cx + dy = q, \quad \text{,,} \quad c \quad \text{,,} \quad d \neq 0 \end{cases}$, can be written in matrix form as $\begin{bmatrix} a & b \\ c & d \end{bmatrix}\begin{bmatrix} x \\ y \end{bmatrix} = \begin{bmatrix} p \\ q \end{bmatrix}$, and the determinant $\Delta = |A| = \begin{vmatrix} a & b \\ c & d \end{vmatrix}$ of the (2×2) matrix $\begin{bmatrix} a & b \\ c & d \end{bmatrix}$ is an alternative notation for the expression $(ad - bc)$.

Section 6.8 involves a (3×3) matrix of the form

$$ax + by + cz = p$$

$$dx + ey + fz = q$$

$$gx + hy + iz = r$$

i.e. three equations in three unknowns which can be written as a square matrix

$$\begin{bmatrix} a & b & c \\ d & e & f \\ g & h & i \end{bmatrix}\begin{bmatrix} x \\ y \\ z \end{bmatrix} = \begin{bmatrix} p \\ q \\ r \end{bmatrix} \text{ and the determinant } \Delta = \begin{vmatrix} a & b & c \\ d & e & f \\ g & h & i \end{vmatrix}$$

expands to

$$\Delta = a\begin{vmatrix} e & f \\ h & i \end{vmatrix} - b\begin{vmatrix} d & f \\ g & i \end{vmatrix} + c\begin{vmatrix} d & e \\ g & h \end{vmatrix}$$

That is, the elements of the (2×2) determinants are obtained by 'covering up' the row and column in which each first row element appears. Determinants of higher order square matrices can be defined in a similar manner. Thus, the matrix notation enables information to be "stored" in rectangular arrays—each corresponding item being stored in a particular position. It is useful when manipulating complete sets of simultaneous (and other) types of equations. Existing, computers can solve, say, a (50×50) matrix in a minute or so.

B.4 LOGARITHMS

If $N = 10^x = \varepsilon^{ax}$, then $10 = \varepsilon^a$ and $1 = a \log_{10} \varepsilon$. Hence, $\log_{10} N = x$ and $\log_\varepsilon N = ax = \dfrac{\log_{10} N}{\log_{10} \varepsilon}$, i.e. $\log_\varepsilon N = \dfrac{\log_{10} N}{0.4343}$. A logarithm being the power to which the base must be raised to give the number.

B.5 TRIGONOMETRY

For a triangle whose lengths of sides are a, b, c, and the included angle opposite c is θ, then

$$c^2 = a^2 + b^2 - 2ab \cos \theta.$$

Alternatives:

$$\sin \theta = -\cos(\theta + \pi/2) = -\sin(\theta + \pi) = -\sin(-\theta)$$

$$\cos \theta = +\sin(\theta + \pi/2) = -\cos(\theta + \pi) = +\cos(-\theta)$$

Expansions:

$$\sin(A + B) = \sin A \cos B + \cos A \sin B$$

$$\cos(A + B) = \cos A \cos B - \sin A \sin B$$

$$\sin 2A = 2 \sin A \cos A \text{ and } \sin^2 A + \cos^2 A = 1$$

$$\cos 2A = \cos^2 A - \sin^2 A = 2\cos^2 A - 1 = 1 - 2\sin^2 A$$

Alternatives:

$$a \sin \theta - b \cos \theta = \surd(a^2 + b^2) \sin(\theta - \phi)$$

$$a \cos \theta + b \sin \theta = \surd(a^2 + b^2) \cos(\theta - \phi),$$

where $\phi = \tan^{-1}(b/a)$

B.6 SERIES

Binomial theorem:

$$(a+b)^n = a^n + na^{n-1}b^1 + \frac{n(n-1)}{1 \times 2}a^{n-6}b^2 + \cdots + b^n.$$

Fourier series:

$$f(t) = \frac{a_0}{2} + \sum_{n=1}^{n=\infty} (a_n \cos n\omega t + b_n \sin n\omega t)$$

where $\omega = \dfrac{2\pi}{\tau}$

$$a_n = \frac{\omega}{\pi}\int_{-\pi/\omega}^{\pi/\omega} f(t) \cos n\omega t dt; b_n = \frac{\omega}{\pi}\int_{-\pi/\omega}^{\pi/\omega} f(t) \sin n\omega t \, dt; \qquad a_0 = \frac{\omega}{\pi}\int_{-\pi/\omega}^{\pi/\omega} f(t) dt,$$

i.e. $\dfrac{a_0}{2}$ = average over periodic $\tau_1 = \dfrac{\alpha\pi}{\omega}$

Maclaurin's theorem:

$$f(x) = f(0) + xf'(0) + \frac{x^2}{2!}f''(0) + \cdots + \frac{x^n}{n!}f^n(0)$$

Hence,

$$\log_\varepsilon(1+x) = x - \frac{x^2}{2} + \frac{x^3}{3} - \frac{x^4}{4} + \cdots$$

$$\varepsilon^x = 1 + x + \frac{x^2}{2!} + \frac{x^3}{3!} + \cdots + \frac{x^n}{n!} (\varepsilon = 2.71828\ldots \text{ incommensurable})$$

$$\sin x = x - \frac{x^3}{3!} + \frac{x^5}{5!} - \cdots = (\varepsilon^{jx} - \varepsilon^{-jx})2j.$$

$$\cos x = 1 - \frac{x^2}{2!} + \frac{x^4}{4!} - \cdots = (\varepsilon^{jx} + \varepsilon^{-rx})/2$$

Therefore, $\varepsilon^{jx} = \cos x + j \sin x$ and $\varepsilon^{-jx} = \cos x - j \sin x$.

B.7 HYPERBOLIC FUNCTIONS

$$\frac{\varepsilon^x - \varepsilon^{-x}}{2} = \sinh x = x + \frac{x^3}{3!} + \frac{x^5}{5!} + \cdots$$

$$\frac{\varepsilon^x + \varepsilon^{-x}}{2} = \cosh x = 1 + \frac{x^2}{2!} + \frac{x^4}{4!} + \cdots$$

$\varepsilon^x = \sinh x + \cosh x$ and $\varepsilon^{-x} = \cosh x - \sinh x$

$\sinh(x + y) = \sinh x \cosh y + \cosh x \sinh y$

$\cosh(x + y) = \cosh x \cosh y + \sinh x \sinh y$

$\cosh^2 x - \sinh^2 x = 1; \cos jx = \cosh x; \sin jx = j \sinh x.$

B.8 PLANE ANGLE MEASUREMENT

Plane angle in analysis is usually expressed either by means of radian measure, degree, multiple or sub-multiple of a complete turn or revolution. These are interconnected (because of their definitions) by numerical factors, namely, $1 \text{ rad} = \left(\frac{180}{\pi}\right)\text{deg} = \left(\frac{1}{2\pi}\right)\text{rev}$. Thus, any plane angle may be expressed as $\theta = |\theta_1|\text{rad} = |\theta_2|\text{deg} = |N|\text{rev}$, where the ratio of the numerics $\frac{|\theta_2|}{|\theta_1|} = \frac{\text{rad}}{\text{deg}} = \frac{180}{\pi} = 57.3$ and $\frac{|\theta_1|}{|N|} = \frac{\text{rev}}{\text{rad}} = 2\pi = 6.28$. Also, $\dot{\theta} = |\omega| \text{ rad/s} = |n| \text{ rev/s}$; hence $|\omega| = 2\pi|n|$ is a numerical identity.

Circular measurement is used in mathematical analysis especially to avoid cumbersome calligraphy and spoiling the elegance of the calculus. Thus, although in general, units or particular modes of measurement (e.g. deg, rev or turn and cycle) only enter into a problem towards the end when actual sizes of physical quantities are involved, an exception to this is plane angle (θ) and its derivatives $\left(\frac{d\theta}{dt} = \omega \text{ and } \frac{d^2\theta}{dt^2} = \alpha\right)$ which, from the outset of mathematical analysis, is usually presupposed to be in radians (i.e. in circular measure). As a consequence, it is implicit in all the usual formulae e.g. $W = T\theta$; $v = r\omega$; $T = I\alpha$; $P = T\omega$; $E = \frac{1}{2}I\omega^2$; $\frac{d^2x}{dt^2} + \omega^2 x = 0$ and $[D^2 + \omega^2]\ \theta = 0$ as used in undamped linear and torsional oscillations (see Examples 0.5, 1.4(ii) and 1.5(iii), etc.).

It may be contended that circular measure of plane angle—being a ratio of like quantities (arc/radius) is completely expressed as a number and, hence, radian need not appear as a unit, e.g. angular velocity and acceleration having units 1/s and 1/s^2, respectively. In other words, plane angle may be regarded as a dimensionless quantity which can be quantified by use of 'radian' as the name of an angle whose circular measure is unity. It is, however, helpful and necessary in quantitative work to insert rad (if not strictly as a unit-symbol) as an indicator of how plane angles have been quantified—i.e. the particular numerals involved have been obtained by circular measure, as is required in equations and formulae which presuppose radian measurement of plane angles.

The logical artifice of replacing rad by unity (i.e. in effect, omitting rad) at any convenient point (usually at the end) in the quantitative or arithmetical stage of a problem is verified in practice by laboratory and full-scale tests, and is irreducibly simple. It also allows of replacing each letter-symbol by its magnitude (number × unit) in physical formulae and in respect of letter-symbols (θ, ω, α) involving plane angles in "radian-restricted" formulae (see Example 0.3), i.e. in which radian is presupposed as the measure of plane angle.

B.9 COMPLEX QUANTITIES

$x + jy$ is a complex quantity in the rectangular (Cartesian x, y) form, where $j = \sqrt{-1}$. Similarly, in the corresponding polar (r, θ) form, $r\cos\theta + j\sin\theta$. The relationship between the two-forms can be shown on a so-called Argand diagram (Fig. B9(a)) in which the horizontal axis records the real part, x, and the vertical axis the imaginary part, y, of the complex number $z = x + jy = r\cos\theta + jr\sin\theta$ for which $x^2 + y^2 = r^2$ and $\tan\theta = y/x$; and r is sometimes called the modulus of z and written $|z|$.

Thus $x + jy$ denotes a line of length $\sqrt{(x^2 + y^2)}$ tilted up at an angle θ to the axis of x such that $\theta = \tan^{-1}(y/x)$. There are, of course, two values of θ between 0 and 2π which have a given tangent (y/x), but only one θ which has a given sine and cosine. Hence, if a line has a length r along the x axis (i.e. $\theta = 0$), then $r(\cos\theta + j\sin\theta)$ denotes a line of the same length, r, tilted up at angle θ. Thus $(\cos\theta + j\sin\theta)$ can be regarded as an operator which turns r counter-clockwise through an angle θ and, because of equivalence (see B6) $\varepsilon^{j\theta}$ is an alternative expression denoting the same operation. Also, $\cos(\omega t + \beta)$ may be regarded as the real part and $\sin(\omega t + \beta)$ the imaginary part of $\varepsilon^{j(\omega t + \beta)}$. The operator $(\cos\theta + j\sin\theta)$ has the value -1 when $\theta = \pi$, i.e. j^2 (which $= -1$) turns the vector r through π rad, and j (which $= \sqrt{-1}$) turns it through $\frac{\pi}{2}$ rad or 90°. In other words, j when repeated upon itself converts $+1$ into -1.

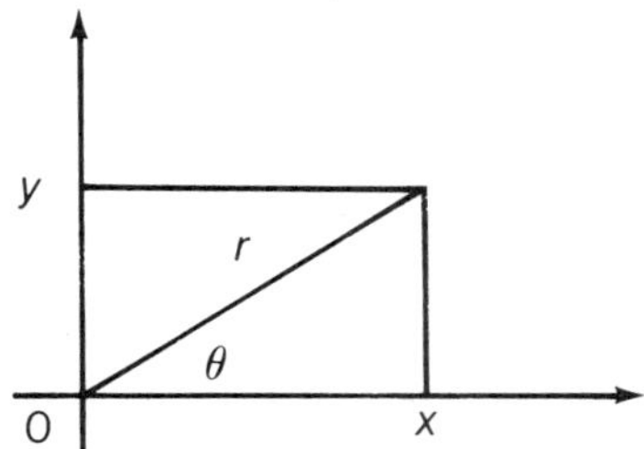

Fig. B9(a) Argand diagram

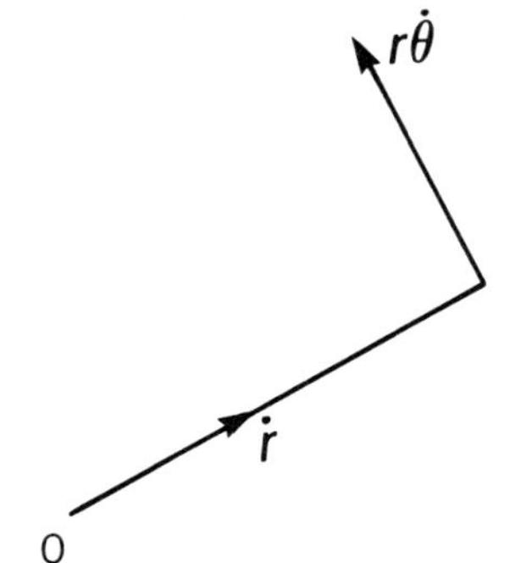

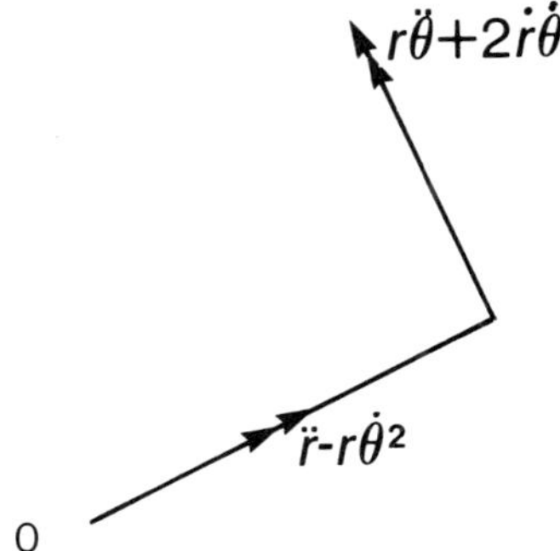

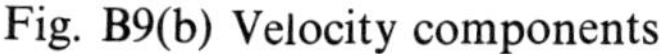
Fig. B9(b) Velocity components

Fig. B9(c) Acceleration components

Thus, referring to Figs. 9(b) and (c), we see that $x = r\varepsilon^{j\theta}$ represents a rotating vector for which $\dot{x} = \dot{r}\varepsilon^{j\theta} + rj\dot{\theta}\varepsilon^{j\theta} = \varepsilon^{j\theta}(\dot{r} + j\dot{\theta})$ and

$$\ddot{x} = \varepsilon^{j\theta}\{\ddot{r} - r\dot{\theta}^2 + j(r\ddot{\theta} + 2\dot{r}\dot{\theta})\}$$

The general solution of the differential equation for s.h.m., namely, $[D^2 + \omega^2]\, x = 0$ is $x = A\varepsilon^{j\omega t} + B\varepsilon^{-j\omega t}$, where A and B are the constants of integration (see B11). Alternatively,

$$\begin{aligned} x &= A(\cos \omega t + j \sin \omega t) + B(\cos \omega t - j \sin \omega t) \\ &= (A + B)\cos \omega t + (A - B)j \sin \omega t \\ &= 2\sqrt{(AB)}(\cos \beta \cos \omega t + \sin \beta \sin \omega t) \\ &= C \cos(\omega t - \beta), \end{aligned}$$

where $C = 2\sqrt{(AB)}$ and $\tan \beta = \dfrac{A - B}{A + B}$.

Also, in contrast to the trigonometrical form, the Fourier series of Sections 1.3 and B6 can be represented in terms of complex numbers, e.g.

$$f(t) = \frac{a_0}{2} + \sum_{n=1}^{n=\infty} (\alpha_n \varepsilon^{jn\omega t} + \alpha_{-n}\varepsilon^{-jn\omega t}),$$

where $2\alpha_n = a_n - jb_n$ and $2\alpha_{-n} = a_n + jb_n$. As in Section 1.3, $\alpha_n\varepsilon^{jn\omega t}$ may be represented as a vector rotating in a counter-clockwise direction with angular velocity $n\omega$, and $\alpha_{-n}\varepsilon^{-jn\omega t}$ in a clockwise direction with angular velocity $-\omega_n$.

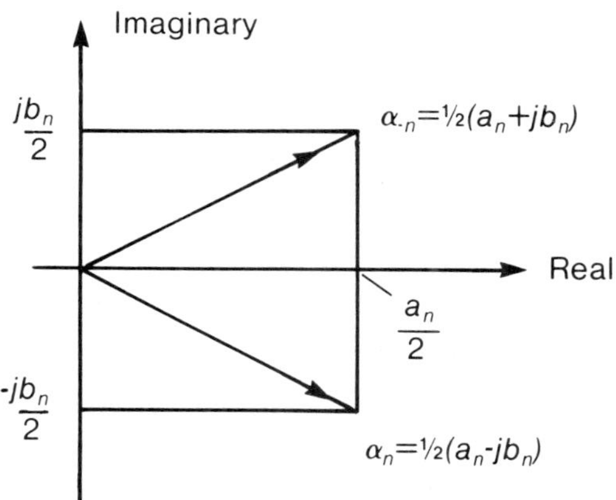

Fig. B9(d) Coefficients of Fourier series

Referring to Fig. B9(d), the imaginary parts of the rotating vectors cancel, and the real parts, being equal, add to double the magnitude such that

$$f(t) = \sum_{n=-\infty}^{n=\infty} \alpha_n \varepsilon^{j\omega t},$$

where $\alpha_n = \frac{\omega}{2\pi} \int_{-\pi/\omega}^{\pi/\omega} f(t)\varepsilon^{-jn\omega t}\,dt$ which includes the constant term $\alpha_0 = \frac{a_0}{2}$

B.10 DIFFERENTIATION

If x and y are functions of t then:

(i) Function of a function rule $\frac{dx}{dt} = \frac{dx}{dy} \cdot \frac{dy}{dt}$—e.g. $\frac{d}{dt} \sin \omega t = \omega \cos \omega t$

(ii) Product: $\frac{d}{dt}(xy) = y\frac{dx}{dt} + x\frac{dy}{dt}$

(iii) Quotient: $\frac{d}{dt}(x/y) = \frac{y\frac{dx}{dt} - x\frac{dy}{dt}}{y^2}$

$$\frac{d}{d\theta} \sin \theta = \cos \theta = \sin\left(\theta + \frac{\pi}{2}\right) = -\cos(\theta + \pi)$$

$$\frac{d}{d\theta} \cos \theta = -\sin \theta = \cos\left(\theta + \frac{\pi}{2}\right) = \sin(\theta + \pi)$$

The symbol D is used to denote the operation of differentiation $\frac{d}{dt}$—e.g.

$\dfrac{d^2x}{dt^2} = D^2x = \ddot{x}$ or, if y is a function of x, D^ny stands for $\dfrac{d^ny}{dx^n}$, where n is any integer.

(iv) Functions of two variables

$$\frac{df(x, y)}{dt} = \frac{\partial f}{\partial x}\cdot\frac{dx}{dt} + \frac{\partial f}{\partial y}\cdot\frac{dy}{dt}.$$

B.11 DIFFERENTIAL EQUATIONS

Differential equations are equations which contain differential coefficients. The *order* of a differential equation is that of the highest differential coefficient; and the equation is *linear* when the independent variable and its derivatives appear only in the first degree and are not multiplied together—e.g. $\dfrac{d^2x}{dt^2} = f(y, t)$ is of the second order and first degree (or linear), where as $\left(\dfrac{dy}{dx}\right)^2 = \phi(x, y)$ is of the first order and second degree.

The general form of linear differential equations of any order with constant coefficients may be written:

$$a_0y_n + a_1y_{n-1} + a_2y_{n-2} + \cdots + a_ny_0 = f(x)$$

where y is a function of x, $y_n = \dfrac{d^ny}{dx^n}$ and the a's are constant and $f(x)$ is a function of x only. This may alternatively, be written: $\phi(D)y = f(x)$, where

$$\phi(D) \equiv a_0D^n + a_1D^{n-1} + \cdots + a_n.$$

The general solution for y must contain n arbitrary constants because the order of the differential equation is n. If $y = v$ is a particular solution (i.e. a particular integral) without arbitrary constants, then $\phi(D)v = f(x)$. Hence, if we put $\dot{y} = v + w$ as the general solution, then

$$\phi(D)v + \phi(D)w = f(x) \quad \text{and} \quad \phi(w) = 0,$$

where w is called the complementary function. Thus, the general solution, $y = w + v$, is the sum of two parts, namely, (i) the complementary function (i.e. when the right hand side of the differential equation is zero) which contains independent constants equal in number to the order of the differential equation, and (ii) the particular integral which is the simplest solution of the differential equation, and does not contain any constants of integration.

We realize that to elucidate some of the mysteries of mechanical vibration requires at least the solution of the second order differential equation with constant coefficients, namely,

$$a\frac{d^2x}{dt^2} + b\frac{dx}{dt} + cx = [aD^2 + bD + c]x = d\cos\omega t.$$

The particular integral gives the forced (steady state) vibration, and the complementary function gives the natural (transient) vibration—the two constants in the latter being decided by the two starting conditions specifying initial displacement and velocity as will be seen in several worked examples in the text.

If in the typical second-order differential equation $[mD^2 + cD + k]x = 0$ for vibrations (as in Section 3.2) the values of m, c, or k are not constant but are, say, functions of x or dx/dt, the equation is 'non-linear', In practice non-linearity occurs in elastic members of built-up structures which usually stiffen (i.e. k increases) as clearances within it are taken up; also, in practical systems, damping is not always viscous, i.e. c is not always constant. Non-linearity makes exact analysis relatively difficult, and recourse is usually necessary to approximate methods or numerical integration. Mass-inductance or mass-capacitance analogies between electrical circuits and vibratory mechanical systems based on the similarity of equations (see Appendix C9) are often used as an aid in predicting the behaviour of complex dynamical systems—the circuit parameters, namely, inductance L, resistance R and capacitance C being almost completely separable. It is relatively easy to achieve oscillatory stability of an electrical circuit by adding inductance or capacitance at certain points in the circuit.

B.12 COMPLEMENTING FUNCTIONS

Particular cases of $a_0D^n + a_1D^{n-1} + \cdots + a_n = f(x)$ required in mechanical vibrations (using D to represent $\frac{d}{dt}$, and x being a function of t) are as follows:

When $n = 1$ we have a linear first order equation with constant coefficients which may be written

$$[D + a]x = 0 = \frac{dx}{dt} + ax$$

Hence, $\int\frac{dx}{x} + a\int dt = \text{constant}$ i.e. $\log_\varepsilon x + at = \log_\varepsilon C$ or $x = C\varepsilon^{-at}$ is the general solution as proved by substitution in the original differential equation. Also, note (e.g. when $n = 2$ and $n = 4$) that ε^{-at} appears as a result of $[D + a]$ and $\varepsilon^{j\omega t}$ as a result of $[D - j\omega]$.

When $n = 2$ we have a linear second order differential equation with constant coefficients which recurs frequently (see Chapter 3) in mechanical vibrations in the form $[D^2 + 2\zeta\omega_n D + \omega_n^2]x = 0$.

Factorizing, we get

$$[D + \omega_n\{\zeta + \sqrt{(\zeta^2 - 1)}\}][D + \omega_n\{\zeta - \sqrt{(\zeta^2 - 1)}\}] = 0$$

or $[D + (a + jb)][D + (a - jb)]x = 0$, where $a = \omega_n\zeta$ and $b = \omega_n\sqrt{(1 - \zeta^2)} = \omega_d$. Hence, the general solution may be expressed as

$$\begin{aligned} x &= C_1\varepsilon^{-(a-jb)t} + C_2\varepsilon^{-(a+jb)t} \\ &= \varepsilon^{-at}(C_1\varepsilon^{jbt} + C_2\varepsilon^{-jbt}) \\ &= \varepsilon^{-at}C\cos(bt - \beta) \\ &= \varepsilon^{-\omega_n\zeta t}C\cos(\omega_n\sqrt{(1 - \zeta^2)}t - \beta), \end{aligned}$$

except when $\zeta = 1$.

The four special cases depend on the numerical values of ζ, namely:

(i) $\zeta = 0$, i.e. $a = 0$ and $b = \omega_n$ or $[D^2 + \omega_n^2]x = 0$ which when factorized as $[D - j\omega_n][D + j\omega_n]x$ yields the general solution $x = C_1\varepsilon^{j\omega_n t} + C_2\varepsilon^{-j\omega_n t}$ which is, alternatively, $x = A\cos\omega_n t + B\sin\omega_n t = C\cos(\omega_n t - \beta)$ or $C\sin(\omega_n t + \alpha)$ as used in Chapters 1, 2 and 4. That the solution of $\ddot{x} + (k/m)x = 0 = [D^2 + \omega_n]x = 0$ is harmonic can be verified by substituting either the trigonometrical or the exponential function in this equation of motion—e.g. for a trial solution let $x = C\varepsilon^{rt}$ from which $\ddot{x} = r^2C\varepsilon^{rt}$ and $r^2 + k/m = 0$ which has two roots $r_{1,2} = \pm\sqrt{(-k/m)} = \pm j\omega_n$. Hence, two integrals of $[D^2 + k/m]x = 0 = m\ddot{x} + kx$ are $x_1 = C_1\varepsilon^{j\omega_n t}$ and $x_2 = C_2\varepsilon^{-j\omega_n t}$. Their sum also satisfies the equation and has two arbitrary constants which is the required number in the solution of a second order differential equation.

(ii) $0 < \zeta < 1$, i.e. $a = \omega_n\zeta$ and $b = \omega_n\sqrt{(1 - \zeta^2)} = \omega_d$ and the solution is $x = C\varepsilon^{-\omega_n\zeta t}\cos(\omega_d t - \beta)$ which is the most usual case in practice—the constants C and β being determined from known values of x and $\dot{x}$ when $t = 0$ as e.g., in Section 3.3 and Example 3.5(i)a.

(iii) $\zeta = 1$, i.e. $a = \omega_n$ and $b = 0$ or $[D + \omega_n]^2x = 0$, the general solution of which is $x = (C_1x + C_2)\varepsilon^{-\omega_n t}$ but this does not follow by substituting $\zeta = 1$ in the formula for the general solution because there would then be only one constant instead of two. The solution is a result of finding function v such that $x = vx^{-\omega_n t}$, i.e. to satisfy the equation $[D + \omega_n]^2v\varepsilon^{-\omega_n t} = 0 = \varepsilon^{-\omega_n t}D^2v$ from which $D^2v = 0$ yielding $v = C_1t + C_2$. Thus, the general solution of $[D + \omega_n]^2x = 0$ is $x = (C_1t + C_2)\varepsilon^{-\omega_n t}$ as used, e.g., in Section 3.4 and Example 3.5(i)b.

(iv) $\zeta > 1$, i.e. $a = \omega_n\zeta$ and $b = j\omega_n\sqrt{(\zeta^2 - 1)} = \omega_d$, and the general solution is that of the previously factorized equation which yields

$$x = \varepsilon^{-\omega_n \zeta t}(C_1 \varepsilon^{\omega_n \sqrt{(\zeta^2 - 1)}t} + C_2 \varepsilon^{-\omega_n \sqrt{(\zeta^2 - 1)}t}).$$

as in Example 3.5(i)c. In these cases we see that, owing to the exponential term $\varepsilon^{-\omega_n \zeta t}$, the complimentary function part of the complete solution dies away as $t \to \infty$ provided that $\zeta \neq 0$.

(v) When n = 4 we have a linear fourth order differential equation with constant coefficients which occurs in analysis of transverse vibration of beams and whirling of shafts in the form $[\mathrm{D}^2 - \lambda^4]u = 0$, where $\lambda^4 = \dfrac{\rho a \omega^2}{EI}$ and $u = \psi(x)$ in $y = u \sin(\omega t + \alpha)$ for free transverse vibration. Factorizing, we get

$$[\mathrm{D}^2 + \lambda^2][\mathrm{D}^2 - \lambda^2]u = 0$$

and

$$[\mathrm{D} - j\lambda][\mathrm{D} + j\lambda][\mathrm{D} - \lambda][\mathrm{D} + \lambda]u = 0$$

Hence,

$$\begin{aligned} u &= a\varepsilon^{j\lambda x} + b\varepsilon^{-j\lambda x} + c\varepsilon^{\lambda x} + d\varepsilon^{-\lambda x} \\ &= (a + b)\cos \lambda x + (a - b)i \sin \lambda x \\ &\quad + (c + d)\cosh \lambda x + (c - d)\sinh \lambda x \end{aligned}$$

i.e. $u = A \cos \lambda x + B \sin \lambda x + C \cosh \lambda x + D \sinh \lambda x$ as used in Section 6.2.

B.13 PARTICULAR INTEGRALS

As a preliminary we may note that $\dfrac{1}{\mathrm{D}}$ denotes integration; for, since

$$\mathrm{D}x = \frac{\mathrm{d}x}{\mathrm{d}t} = f(t), \quad \text{then } x = \frac{1}{\mathrm{D}} f(t) = \int f(t)\mathrm{d}t.$$

In mechanical vibrations the function $f(t)$ in linear second order differential equations with constant coefficients, namely, $\phi(\mathrm{D})x = f(t)$, is usually harmonic as in the analysis of forced vibrations of Chapters 4 and 5. The particular integral, $x = \dfrac{1}{\phi(\mathrm{D})} f(t)$, can be found with the help of certain easily-proved theorems such as:

(i) if $f(t)$ is exponential expressed as $\varepsilon^{\omega t}$, then

$$x = \frac{1}{\phi(\mathrm{D})}\varepsilon^{\omega t} = \varepsilon^{\omega t}\frac{1}{\phi(\omega)} \qquad (\text{provided } \phi(\omega) \neq 0)$$

(ii) if $f(t) = \varepsilon^{\omega t} V$, where V is either a constant or a function of t, then

$$x = \frac{1}{\phi(\mathrm{D})}\{\varepsilon^{\omega t} V\} = \varepsilon^{\omega t}\frac{1}{\phi(\mathrm{D} + \omega)} V$$

i.e. $\varepsilon^{\omega t}$ can be moved to the left of the operator function $\phi(\mathrm{D})$ provided that D is replaced by $(\mathrm{D}+a)$

(iii) if $f(t)$ is trigonometrical—e.g. $f(t)=V\cos(\omega t+\beta)$ then $x=\dfrac{1}{\phi(\mathrm{D})}$ $V\cos(\omega t+\beta)$ or the real part of $\dfrac{1}{\phi(\mathrm{D})}V\varepsilon^{j(\omega t+\beta)}=(\mathrm{R})\varepsilon^{j(\omega t+\beta)}\dfrac{1}{\phi(\mathrm{D}+j\omega)}V$ and, as a corollary, if $f(t)=\cos(\omega t+\beta)$ and $\phi(\mathrm{D})$ is an even function of D – say, $\psi(\mathrm{D}^2)$ then, since $\mathrm{D}^2\cos(\omega t+\beta)=-\omega^2\cos(\omega t+\beta)$, $\phi(\mathrm{D})=\psi(\mathrm{D}^2)=\psi(-\omega^2)$ and

$$x=\frac{1}{\phi(\mathrm{D})}\cos(\omega t+\beta)=\frac{1}{\psi(\mathrm{D}^2)}\cos(\omega t+\beta)=\frac{\cos(\omega t+\beta)}{\psi(-\omega^2)}$$

as used in Section 4.1.

Particular integrals found via these theorems can, of course, be double-checked by substitution to make sure they satisfy the original differential equations.

In the case of **Example 4.1 (ii)**, $[\mathrm{D}^2+\omega_n^2]x=\omega_n^2\,a\cos\omega t$ the particular integral is

$$x=\frac{1}{[\mathrm{D}^2+\omega_n^2]}\omega_n^2 a\cos\omega t=\frac{\omega_n^2}{\omega_n^2-\omega^2}a\cos\omega t$$

(by substituting $-\omega^2$ for D^2) provided $\omega\neq\omega_n$. Resonance occurs when $\omega=\omega_n$, and the particular integral is found as the real part of the exponential $\omega_n^2 a\varepsilon^{j\omega_n t}$ namely, $x=(\mathrm{R})\left[\dfrac{\omega_n^2 a}{(\mathrm{D}+j\omega_n)(\mathrm{D}-j\omega_n)}\right]\varepsilon^{j\omega_n t}$. Hence, putting $\mathrm{D}=j\omega_n$ in the factor which does not vanish we get $2j\omega_n$ and then moving $\varepsilon^{j\omega t}$ to the left of operator $(\mathrm{D}-j\omega_n)$ by replacing D by $(\mathrm{D}+j\omega)$ we get

$$x=(\mathrm{R})\frac{\omega_n^2 a\varepsilon^{j\omega_n t}}{2j\omega_n}\frac{1}{\mathrm{D}}=(\mathrm{R})\frac{\omega_n^2 at\varepsilon^{j\omega_n t}}{2j\omega_n}$$

where $\dfrac{1}{\mathrm{D}}$ denotes the integral of 1, namely, t.

Thus, $x=(\mathrm{R})\omega_n^2 at\left(\dfrac{\cos\omega_n t+j\sin\omega_n t}{2j\omega_n}\right)$ i.e. the particular integral is $x=\dfrac{\omega_n}{2}at\sin\omega_n t$ which when substituted in $[\mathrm{D}^2+\omega_n^2]x$ gives $\omega_n^2\,a\cos\omega_n t$, as required.

In the case of **Example 4.4(iii)**, $\phi(\mathrm{D})x = [\mathrm{D}^2 + 2\zeta\omega_n + \omega_n^2]x = (F/m)\cos\omega t$ for which the particular integral is the real part of $(F/m)\varepsilon^{j\omega t}$ i.e.

$$
\begin{aligned}
(\mathrm{R})\frac{F}{m}\frac{1}{\phi(\mathrm{D})}\varepsilon^{j\omega t} &= (\mathrm{R})\frac{F}{m}\varepsilon^{j\omega t}\frac{1}{\phi(\mathrm{D}+j\omega)} \\
&= (\mathrm{R})\frac{F}{m}\varepsilon^{j\omega t}\frac{1}{-\omega^2 + 2\zeta\omega_n j\omega + \omega_n^2} \\
&= (\mathrm{R})\frac{F}{m}\frac{\varepsilon^{j\omega t}}{(\omega_n^2 - \omega^2 + 2\zeta\omega_n j\omega)} \\
&= (\mathrm{R})\frac{F}{m}\varepsilon^{j\omega t}\left\{\frac{(\omega_n^2 - \omega^2)^2 - 2\zeta\omega_n j\omega}{(\omega_n^2 - \omega^2)^2 + (2\zeta\omega_n\omega)^2}\right\} \\
&= \frac{F}{m}\left\{\frac{(\omega_n^2 - \omega^2)^2\cos\omega t + 2\zeta\omega_n\omega\sin\omega t}{(\omega_n^2 - \omega^2)^2 + (2\zeta\omega_n\omega)^2}\right\} \\
&= \frac{F}{m}\left\{\frac{\cos\beta\cos\omega t + \sin\beta\sin\omega t}{\sqrt{\{(\omega_n^2 - \omega^2)^2 + (2\zeta\omega_n\omega)^2\}}}\right\} \\
&= \frac{F}{m}\frac{\cos(\omega t - \beta)}{\sqrt{\{(\omega_n^2 - \omega^2)^2 + (2\zeta\omega_n\omega)^2\}}},
\end{aligned}
$$

where

$$
\beta = \tan^{-1}\left(\frac{2\zeta\omega_n\omega}{\omega_n^2 - \omega^2}\right).
$$

Appendix C

Detection and measurement of vibration and noise

Measurement of vibration in all its aspects e.g. acceleration, velocity, amplitude and relative displacement, transmitted forces and stress, energy dissipation, noise, etc., is a field in itself covered by instrumentation specialists, as will be appreciated from the following statements—necessarily brief, because this book is not a primer on instrumentation, as such.

C.1 SENSITIVITY OF MEASURING INSTRUMENTS

In general, measuring instruments are either of the indicating or balancing type. In the former, a pointer is usually moved by an impressed effect against a resistance such as a spring or pressure gauge. In the latter type the impressed effect is nullified by a known counter-effect to give zero deflection of the indicating element such as an electrical bridge. If x_0 is the correct value of a measurement recorded as x, then $\delta x = x_0 - x$ is the absolute error, and $\delta x/x$ is the relative error. The sensitivity of an instrument is the change in reading per unit change in the measured quantity, e.g. the sensitivity of a spring-balance may be quoted in mm/kg and that of an electrical strain-gauge in fractional change of resistance per unit of strain.

If an instrument is being used to measure a quantity which suddenly changes its value, e.g. a galvanometer when current is switched on, the indicating pointer oscillates at the natural frequency $\left(\frac{\omega_n}{\text{rad}} = \sqrt{(k/m)}\text{—see Example 0.5 and Section 2.2}\right)$ of the instrument, and the die-away rate is governed by its damping. The same thing happens if the instrument is given a jolt (see Section 4.2). An instrument having a light spring control (i.e. k is small) will have a low undamped natural frequency (ω_n) but is incapable of faithfully recording a rapidly-changing quantity; and a low-frequency instrument is susceptible to extraneous disturbances. If, however, the instrument is correctly damped it will record the mean value of a fluctuating quantity.

Since $\omega_n/\text{rad} = \sqrt{(k/m)}$ it follows that to obtain a sensitive instrument for measuring fluctuating quantities (including the effect of jolts in, say, moving vehicles), it is necessary that its natural frequency should be high, which requires m to be relatively small. This is most effectively done by the use of an electrical pick-up or transducer in conjunction with an amplifier and indicating-meter such as an oscilloscope whose electron beam has negligible inertia and, hence has an extremely rapid response.

C.2 TRANSDUCERS IN INSTRUMENTATION

'Transducer' is the name given to any device which converts one physical quantity, e.g. a vibration of small amplitude, into another more-easily usable (say, an electrical) quantity either for measurements, control or operation of processes. For measurement purposes it is usual to transduce into electrical voltage or current and to record what is happening, e.g. amplified as a visual image on the screen of a cathode-ray oscilloscope which is especially useful for the display of qualitative results. Electrical signals can also be processed, e.g. integrated, differentiated, selectively filtered, immediately recorded and presented in analogue or digital form for use in a computer. Electrical filters can suppress the frequencies in which engineers are not interested; or, alternatively, the graphical recording of a total vibration on magnetic tape can be filtered. One of the effects most used for changing a mechanical quantity into a corresponding electrical one is the change in electrical resistance when a wire is stretched; another is the piezo-electric effect, namely, when certain crystals are deformed, a voltage is produced between their surfaces. Other effects known to physics are used for the transduction of mechanical effects into their electrical analogies or equivalents, e.g. thermo-electric voltages produced when a junction between dissimilar metals is heated, are used for temperature measurement; and changes in capacitance or induction when coils carrying alternating currents, or plates fed with direct current move relative to each other, are used for measuring displacement and pressure. Thus, the range of electrical transducers is extensive for measurement of physical quantities including displacement, velocity, pressure, flow rate, viscosity, temperature, etc. In particular, as resistance strain-gauge (which may form one arm of a Wheatstone bridge) is a 'pick-up' in which the electrical resistance of a zig-zag of fine wire, etchedfoil or filament of semi-conductor material changes as it is stretched when glued or cemented to, but insulated from, the part under investigation because, say, of a suspicion of its possible failure by fatigue. The appearance of a crack in a component reveals the direction of the principal stress and, hence, the correct position of the axis of the strain-gauge in investigation of a new component is at right angles to the crack on the one to be replaced.

To record what is happening is, therefore, relatively easy since amplification is electrically straightforward, and recording apparatus can be placed away from the vibrating system. For automatic control purposes, however, it may be preferable to transduce a measurement into fluidic terms capable of being applied directly to the operation of a mechanism being controlled. Transducers are, of course, only part of a complete instrumentation system, the integrity of which is very important since unless the initial transfer from mechanical to electrical effect is accurate, all else is useless for quantitative and evaluation purposes. Hence, it is important to make the choice of transducer in accordance with the type of problem being investigated, and range of vibration involved. Care is needed with calibration—e.g. the inclusion of connecting wires in the calibration of pick-ups. Also, unless mains and leads are screened, magnetic fields of currents flowing may cause interference and distortion of a genuine signal being transmitted.

The voltage waveform which emanates from vibration pick-ups can be amplified for display on a cathode-ray screen across which the spot can be adjusted to trace one complete cycle, and continually repeat it. A predominating waveform envelope associated with the period of rotation of a mechanism may be revealed containing several higher-order frequency components caused by the vibrations of certain parts composing the mechanism. By increasing the time base frequency—i.e. the rate at which the spot traverses the screen, a picture of these higher-order components can be obtained.

C.3 SEISMIC PICK-UPS

The earliest displacement pick-ups (referred to as 'seismic'; also see Section 4.2) were mechanical and designed for recording earth tremors and transient shocks for which they are still used (see Examples 5.3(iii) and (iv)) as distinct from pick-ups which give an electrical output. Certain natural crystals such as quartz, Rochelle salt and some semi-conductors show a piezo-electric effect which produces a voltage on opposite faces when such crystals are deformed or strained and, in effect, act mainly as condensers. Lead zirconate has a greater sensitivity than quartz for the same frequency, but is limited to 250°C as against 350°C for quartz when attached, say, to hot engine parts. These types of pick-up have a desirably high resonant frequency, e.g. from 30 to 100 kHz, and can be compacted within a 1 cm^3. In principle, pick-ups which are piezo-electric in function consist of a fairly small mass held firmly or cemented to a crystal as shown diagrammatically in Fig. C.3(a).

Another type is the seismic moving coil pick-up shown in Fig. C.3(b) basically comprising a mass mounted such that its deflection is linear with the imposed displacement, and the electrical output depends on the rate of cutting

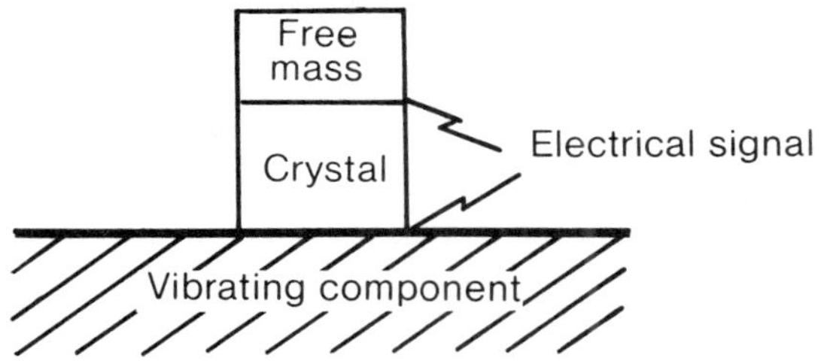

Fig. C3(a) Prezo-electric pick-up

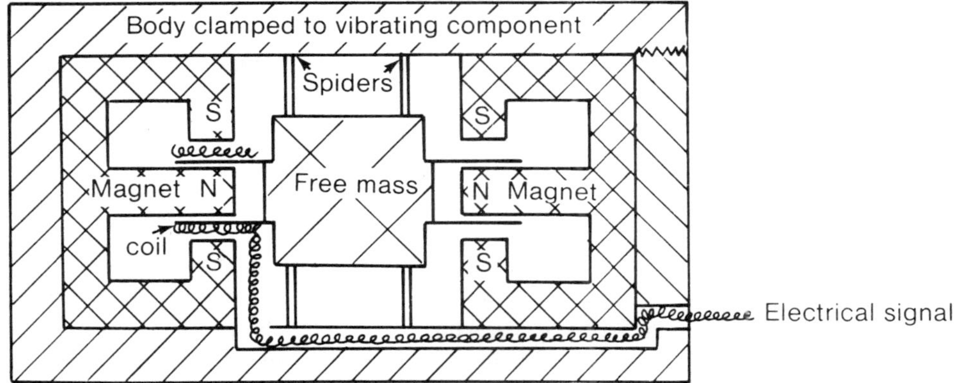

Fig. C3(b) Seismic moving-coil pick-up

of lines of force from the magnets—i.e. the output is velocity responsive which requires a single integration to give displacement or a single differentiation to give acceleration. Moving coil types are used at frequencies from 25 Hz upto 2kHz. Seismic pick-ups comprising a mass mounted on springs are dealt with in detail in Sections 4.2 and 5.3 where it is shown that when required to measure displacement its design must be such that its resonant frequency $\omega_n = \sqrt{(k/m)}$ must be lower than the frequency to be measured i.e. 'soft' springs and relatively large mass, and when required to measure acceleration its resonant frequency must be higher than that of the frequency to be measured. Seismic pick-ups read movements relative to the earth; hence, to obtain movement or acceleration relative to any other part of the system it is necessary to combine the outputs of two such pick-ups vectorially. Also, all seismic pick-ups are sensitive to side effects, e.g. to movements transverse to their axis, and from torsional oscillation which may swivel a moving coil in the magnetic field.

C.4 CALIBRATION AND WAVE SELECTION

Thus, vibration pick-ups of small mass compared with that of the part of a mechanism on which they are carefully mounted, together with the use of suitable electrical equipment (e.g. amplifier, integrator, filter and indicating—meter) to deal with their output voltages can be made to detect and measure any of the variables—displacement, velocity and acceleration. Also, by means of, say, an electro-magnetic sinusoidal vibrator whose amplitude can be measured accurately by means of a microscope, pick-ups can be calibrated in terms of known displacements, velocities, accelerations and frequencies, e.g. a pick-up may be specified as producing so many millivolts per μm of displacement. Electrical filters may be used to select the frequency band of particular interest—i.e. allow the voltages of only certain frequencies to pass through for measurement; and a wave-analyser is an instrument by means of which the frequency components of a given wave-form can be separately examined—i.e. it enables a spectrum analysis to be made of a response in terms of its individual frequency components analogous to that of the separation of white light into separate colours.

C.5 NOISE LEVELS

We may note that a vibrating body works against the impedance of the air and, hence, acoustical power is radiated from it—the sound radiation being nearly proportional to the acceleration. Noise levels may be measured or compared by using a microphone (i.e. an air-pressure instrument), and the waveform and frequency of the noise recorded. Also, we may marvel at the human ear, which is a remarkably sensitive and discriminating vibration-detecting device on line to the computer of the brain—the audible range for humans being approximately 18 Hz to 18 kHz. It is possible to use a doctor's stethoscope on external parts of, say, a running engine—the ear being a good detector of resonance, although other aids have to be used to determine which particular mode is being excited.

Many metal-working operations are inherently noisy, e.g. the shearing of metal to a desired shape, and the die-punching of holes, etc. The more noise a machine makes implies an increase in vibrations and consequent undesirable stresses and strains in parts of the mechanism.

C.6 CHOICE OF PICK-UPS

One of the difficulties of measuring vibrations in, say, an engine, mechanism or framework is the absence of a fixed reference point, and another is that of leading the signals out of machinery—particularly from rotating parts. Seismic pick-ups are usually used for monitoring the vibration of housings and bearings, etc; and accelerometers where, say, local resonance causes excessive wear, whereas fatigue-cracks are usually investigated with strain-gauges. Pulsations in pipes conveying liquids may be investigated by means of pressure transducers or foil strain-gauges fitted circumferentially and thus, in effect, making the pipe into a pressure transducer.

C.7 NOISE AND THE DECIBEL

Because vibrations propagate through elastic media in the form of waves, an important quantity characterizing the magnitude of a vibration and a wave is its root mean square (r.m.s.) value. Although the perception of sound by the human ear is a complex process, it is generally assumed that the ear responds to the sound energy as averaged over a certain time, and the quantity usually measured is the r.m.s. value of sound pressure. The range perceived as sound is very large, e.g. at 1 kHz the weakest sound pressure detected by the human ear (i.e. at threshold of hearing) is 0.0002 μbar (2×10^{-5} N/m^2), whereas the largest perceived without pain is about 1 mbar (100 N/m^2). Thus, this scale of sound pressure covers the range $0.2{:}10^6$ and because of this, a logarithmic scale is used in defining the decibel (dB) as ten times the logarithm of the ratio between two quantities of power. Thus, because sound power is related to the square of sound pressure, a convenient (logarithmic and relative) scale for noise measurement is given by

$$\frac{\text{sound pressure level}}{\text{dB}} = 10\log_{10}\left(\frac{p^2}{p_0^2}\right) = 20\log_{10}\left(\frac{p}{p_0}\right),$$

where p is the sound pressure being measured and p_0 is a reference pressure (usually taken as 0.002 2μbar), namely, the base above which the level of the given sound pressure is measured, e.g. zero and 20 dB corresponds to sound pressure ratios of 1 and 10, respectively. It follows that the threshold of pain occurs when the sound pressure approaches 1 mbar which corresponds to 134 dB, e.g. conversational speech has a sound level of about 60 dB corresponding to a sound pressure of 0.2 μbar, whereas a pneumatic road-drill creates a noise of about 120 to 130 dB. Thus, by use of the decibel, a scale of sound pressure of 1 to 10 is transformed into a scale of 0 to 120 dB which is a more convenient scale to use.

EXAMPLE C7—illustrative of the interconnection between sound pressure and the decibel.

(i) If the sound pressure is found to be 3.56 μbar, what is the sound pressure level in dB relative to the usual reference pressure of 0.0002 μbar?

$$\frac{p}{p_0} = \frac{3.56}{0.0002} = 1.78 \times 10^4$$

Hence, $\dfrac{\text{sound pressure level}}{\text{dB}} = 20 \times 4 \log_{10} 1.78 = 85$

(ii) If the sound pressure level is 74 dB relative to 0.0002 μbar, what is the actual sound pressure?

$74 = 20 \log_{10}\left(\dfrac{p}{p_0}\right)$. Hence $p_0 = \text{antilog } 3.7 = 5012$ and $p = 0.1 \text{ N/m}^2$.

C.8 NOISE EVALUATION AND CONTROL CRITERIA

The perception of sound by the human ear consists of processes such that no simple or unique relationship exists between the physical measurement of a sound pressure level and perception, e.g. the loudness of a certain pure tone (or individual harmonic) may be judged to sound different from that of another pure tone or combination of tones, even if the sound pressure level is the same in all cases. Thus, if a number of pure tones are combined into a complex sound, not only the loudness and pitch determine the human perception of the sound, but there is a third factor, the timbre, which depends on the harmonic content of the sound and its transient behaviour. Even though many details of hearing are still not well understood, the concept of 'loudness', as predicable from the measurement of sound pressure levels, has been fairly well agreed upon; but the concepts of annoyance, noisiness and noise pollution level have not yet been so firmly established. Certain aspects of annoyance are connected with loudness; and psycho-acoustical experiments indicate that high-pitched (high-frequency) noise is more annoying than low-pitched noise—the 'high' frequencies here ranging from around 1.5 kHz to, say, 10 kHz. Also, if noise is intermittent or irregular, or even rhythmic, it may be considerably more annoying than a 'steady' noise of the same loudness. At present, it seems that noise criteria relate (i) to risk of damage to hearing—based on physical measurements by sound level meters, and (ii) to annoyance (e.g. causing speech interference)—based on frequency analysis techniques.

For suggestions on further reading and study, see Appendix D, (12) and (13).

C.9 MECHANICAL–ELECTRICAL ANALOGIES

Where physical systems are so complex that a mathematical or analytical solution is impossible, it is often possible to devise analogous electrical circuits on which experiments can be made—an analogy being a relationship of mutual similarity appearing in two or more fields of knowledge. Thus, the identification of an analogy facilitates the transfer of knowledge and procedures of analysis of behaviour from one field of study to another. Such a relationship exists between vibrating mechanical systems and electrical networks based on the similarity of differential equations, e.g. referring to Fig. C9(a), which represents the essentials of damped forced oscillatory or linear

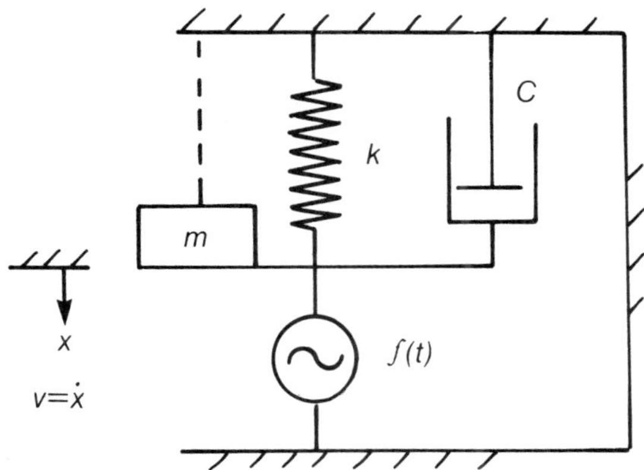

Fig. C9(a) Mechanical damped forced vibratory linear system

vibratory mechanical systems, the general differential equation of motion may be written:

$$[mD^2 + cD + k]x = f(t)$$

or

$$m\dot{v} + cv + k\int v\mathrm{d}t = f(t).$$

This has electrical equivalents:
(i) on the mass-capacitance (or force-current) analogy (Fig. C9(b)) because

$$\left[CD^2 + \frac{1}{R}D + \frac{1}{L}\right]V = \frac{\mathrm{d}I(t)}{\mathrm{d}t}$$

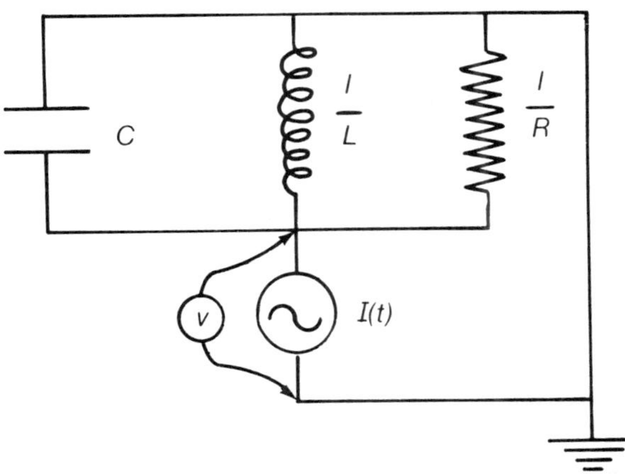

Fig. C9(b) Electrical mass-capacitance (or force-current) analogy of Fig. 4.6(c)

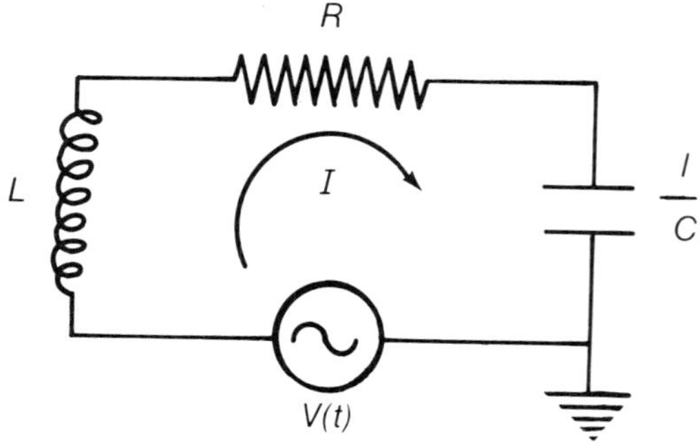

Fig. C9(c) Electrical mass-inductance (or force-voltage) analogy of Fig. 4.6(c)

or

$$C\dot{V} + \frac{1}{R}V + \frac{1}{L}\int V\mathrm{d}t = I(t)$$

and (ii) on the mass-inductance (or force-voltage) analogy (Fig C9(c)) because

$$\left[L\mathrm{D}^2 + R\mathrm{D} + \frac{1}{C}\right]Q = V(t)$$

or

$$L\dot{I} + RI + \frac{1}{C}\int I\mathrm{d}t = V(t)$$

These three sets of equations are similar in form and differ only in the letter-symbols used as tabulated:

Mechanical			Electrical Equivalents			
Physical quantity	Linear	Torsional	mass-capacitance analogy		mass-inductance analogy	
Mass or inertia	m	I	C	capacitance	L	inductance
Force or torque	f	T	I	current	V	voltage
Compliance	$1/k$	$1/k$	L	inductance	C	capacitance
Damping	c	c	$1/R$	conductance	R	resistance
Velocity	v	$\dot{\theta}$	V	voltage	I	current
Displacement	x	θ	$\int V\mathrm{d}t$	—	Q	charge $\int T\mathrm{d}t$

Thus, on the mass-capacitance and mass-inductance analogies, the force causing elastic deformation, kx or $k\int v\mathrm{d}t$, would be $I = \frac{1}{L}\int V\mathrm{d}t$ and $V = \frac{1}{C}\int I\mathrm{d}t = \frac{Q}{C}$, respectively; kinetic energy $\frac{1}{2}mv^2$ would be $\frac{1}{2}CV^2$ and $\frac{1}{2}LI^2$, respectively; strain energy $\frac{1}{2}LI^2$ and $\frac{1}{2}CV^2$, respectively; power dissipated by damping cv^2 would be VI or V^2/R and RI^2, respectively. The mass-capacitance analogy is preferred, because the line-diagrams of mechanical systems (Fig. C9(a)) and the analogous electrical circuits (Fig. C9(b)) can be drawn so that they 'match', whereas on the mass-industance analogy they cannot because, although the differential equations have the same form, mechanical components in parallel (Fig. C9(a)) become electrical components in series (Fig. C9(c)), and vice versa.

Appendix D

Books suggested for further reading and study

(1) J. L. Meriam, *Dynamics* (John Wiley & Sons, 1975).
(2) H. C. Hibbeler, *Engineering Mechanics: Dynamics* (Collier-Macmillan, 1978).
(3) J. L. M. Morrison and B. Crossland, *An Introduction to the Mechanics of Machines* (Longman, 1964).
(4) J. M. Prentis and F. A. Leckie, *Mechanical Vibrations: An Introduction to Matrix Methods* (Longman, 1963).
(5) S. Timeskenko and W. Weaver, *Vibration Problems in Engineering* (McGraw-Hill, 1974). Contains computer programs.
(6) C. F. Beards, *Vibration Analysis and Control System Dynamics* (Ellis Harwood, 1981).
(7) L. Mewovitch, *Elements of Vibration Analysis* (McGraw-Hill, 1975). Contains an introduction to finite element methods.
(8) W. T. Thomson, *Theory of Vibrations with Applications* (George Allen & Unwin, 1983).
(9) K. A. Stroud, *Fourier Series and Harmonic Analysis* (Stanley Thomas, 1984).
(10) Tonia Cope, *Computing using Basic: An Interactive Approach* (Ellis Horwood, 1981).
(11) Roy Atherton, *Structured Programming with BBC BASIC* (Ellis Horwood, 1983).
For those with a BBC micro-computer.
(12) R. G. White and J. G. Walker, *Noise and Vibration* (Ellis Horwood, 1982). A comprehensive review.
(13) Technical literature booklets issued by Bruel and Kjaer, Naerum, Denmark on
(a) Acoustic Noise Measurements, and
(b) Mechanical Vibration and Shock Measurements.

Answers to exercises

Chapter 1

(1) $x = 9.42 \sin(\omega t + 72°51')$, (2) $11.37\varepsilon^{j\pi/3.44}$ (3) 3.17 mm. (4) $4.2 \sin(5t + \beta - 30°)$ and $\sin(5t + \beta - 80°)$ (5) 15h 10 min; 12.55 mm/min. (6) 4.07 kg m^2 or 4.07 Nms^2. (7) 2.56 m.

Chapter 2

(2) 1.5 kg, 0.134 N/mm. (3) 10/s. (4) 1.9s. (6) 176 mm. (8) $\sqrt{\left(\frac{ga}{Lh}\right)}$. (9) $24g$. (10) 2.84 N/mm; 0.4s. (11) $\omega_n = \sqrt{\frac{2GJ}{LI}}$. (12) $X_1/X_2 = 0.414$ and -2.414. (13) $\tau_n = 2\pi\sqrt{(m/\rho g A)}$.

Chapter 3

(1) 0.83. (2) $\frac{c}{2\sqrt{(IR)}}$. (3) $x/x_0 = 1.02\varepsilon^{-0.2\omega_n t}\sin(0.98\omega_n t + 78°28')$; $x/x_0 = (1 + \omega_n t)\varepsilon^{-\omega_n t}$; $x/x_0 = 1.08\varepsilon^{-0.268\omega_n t} - 0.0774\varepsilon^{-3.73\omega_n t}$.

(4) (a) $\zeta = 0$, t	0.1	0.2	0.3	0.4	0.5	0.6	0.7	0.8	0.9	1.0
x	81	31	−31	−81	−100	−81	−31	31	81	100etc.

$\zeta = 0.1$, t	0.2	0.4	0.6	0.8	1.0	1.2	1.4	1.6
x	40	−67	−53	23.2	53.7	11.7	−35.5	−28.6

$\zeta = 0.1$, t	1.8	2.0	2.2	2.4	2.6	2.8	3.0
x	11.2	28.6	6.8	−18.4	−15.7	5.6	15.2

$\zeta - 1.0$, t	0.25	0.5	0.75	1.0	1.25	1.5
x	53.4	18	5.1	1.35	0.34	0.08

$\zeta = 5.0$, t	0.2	0.4	0.6	0.8	1.0	1.5	2.0	2.5	3.0
x	83	71.5	62.9	55.5	49	36	26	19	13.9

(b) 27 per cent, (c) 2.2 s (5) $\delta = \log(x_1/x_2) = 2\pi\zeta/\sqrt{(1-\zeta^2)}$. (6) (a) 28.3 m/s; (b) 25.2 kNs/m; (c) 0.214 s (7) 4.77 cycle/s; 0.022; 0.0035; 0.42 Ns/m. (8) 0.399; 0.0635; 5 cycle/s. (9) 0.10; 3.17 cycle/s; 0.631; 1.88. (10) 11s. (11) 242 mm; 0.215; 2.38s; 2.36×10^4 kg mm^2.

Chapter 4

(1) $\dfrac{z}{X} = \dfrac{\omega^2}{\omega_n^2 - \omega^2}$. (2) 0.0637 Ns/mm. (3) 21.7 mm; 1.974 N/mm. (4)(a) 2.28 Hz; 9.3 mm; 90°, (b) 1.533 Hz; 11.22 mm; 40°66′. (5) 0.999; 0.990; 0.960; 0.707. (6) 0.0667. (7) 10.77 N/mm. (8) $\omega_n = 40$ rad/s; (a) $\frac{1}{24}$; $\frac{1}{99}$; (b) $\frac{1}{17}$; $\frac{1}{44}$. (9) (i) 0.79 mm; 172°(ii) 1/5.58. (iii) 1.42 kN; 120°.

Chapter 5

(1) 9.9 rad/s = 1.575 Hz; 79.5 km/h; 147 mm; 75.8 mm. (3) (a) 0.11; 100 cycles/min (b) 2.49; 26°6′ (4) (a) 0.00606 (b) 3.26 mm; 0.364 s.

Chapter 6

(1) $\omega = \lambda^2 \sqrt{\left(\dfrac{EI}{\rho a}\right)}$, where tanh $\lambda l = \tan \lambda l$ or $\lambda l = 3.93$, 7.07, etc.

(2) $\omega = \lambda^2 \sqrt{\left(\dfrac{EIl}{M_b}\right)}$, where λ is determined from

$$(1 + \cosh \lambda l \cos \lambda l) = ml\frac{M_o}{M_b}(\sin \lambda l \cosh \lambda l - \cos \lambda l \cosh \lambda l)$$

(3) $\omega = \lambda^2 \sqrt{\left(\dfrac{EI}{\rho a}\right)}$ and $\lambda l = 0$, 4.73 from $\cosh \lambda l \cos \lambda l = 1$. Nodes for lowest frequency $0.244l$ from ends.

(4) 19 GN/m^2. (5) 175 Hz. (6) 19.3 cycles/s. (7) (i) $\omega_1 = 5.5\sqrt{\left(\dfrac{EI}{ml^3}\right)}$; 3 cycles/s or 3 Hz (ii) $\dfrac{1}{\omega^2} = \dfrac{ml^3}{30.375\, EI} + \left(\dfrac{l}{\pi}\right)^4 \dfrac{16\rho}{Ed^2}$; 2.92 Hz (8) 505 Hz; 6.27 MN/mm.

Index